How To Write, Publish & Sell Your Own Aviation Books

CHEVY ALDEN

Foreword by Walter J. Boyne

TRI-PACER PRESS
Pembroke Pines, Florida

Published by:
Tri-Pacer Press
P.O. Box 840111
Pembroke Pines, FL 33084-2111

Library of Congress Catalog Card Number: 97-90914
ISBN: 0-9619991-3-6

First Printing: March 1998

Printed in the United States of America

DISCLAIMER: This book is a factual account of the author's experiences with writing and publishing. It is sold with the understanding that the author and publisher are not engaged in rendering legal, accounting or other professional services. If legal or other expert assistance is required, the services of a competent professional should be sought. This book is also sold with the understanding that self-publishing is not a get-rich scheme. Anyone who decides to write and publish a book must expect to invest a lot of time and effort. For many people, self-publishing is more lucrative than selling manuscripts to another publisher, and many have built solid, growing, rewarding businesses. Every effort has been made to make this book as complete and as accurate as possible. However, there may be mistakes both typographical and in content. Therefore, this text should be used only as a general guide and not as the ultimate source of writing and publishing information. Furthermore, this book contains information on writing and publishing only up to the printing date. The purpose of this book is to educate and entertain. The author and publisher shall have neither liability nor responsibility to any person or entity with respect to any loss or damage caused, or alleged to be caused, directly or indirectly by the information contained in this book.

In memory of Ernest K. Gann...author, pilot, sailor!

Books by Chevy Alden

Black Falcon
How To Get Published - Guaranteed
How To Write, Publish & Sell Your Own Aviation Books

Future Books by Chevy Alden

Avenge
Death Wreath
How To Promote & Market Your Own Books
How To Write, Publish & Sell Your Own Nautical Books
Mountains In The Sky: True Tales Of A Sport Pilot
Unplugged: An Autobiographical Odyssey Through One Author's Life

ACKNOWLEDGMENTS

A "champagne toast" goes out to the eighty-four publishers who rejected my first book! Without the dozens of rejection letters I received for "Black Falcon", I never would have explored the wonders of self-publishing! My life will never be the same!

A "special thank you" goes out to Walter J. Boyne for giving my aviation writing career an astronomical boost in 1996!

A "warm thank you" goes out to all those individuals who helped advance my writing/publishing career during the many years I've been putting pen to paper!

ATTENTION: AVIATION ORGANIZATIONS

Is your organization looking for a highly-motivated speaker to talk about writing and publishing aviation books? Chevy Alden makes year-round guest appearances at aviation trade shows, meetings, conventions and gatherings. For detailed information about scheduling Chevy Alden, please contact:

Alden Speaking Engagements
c/o Tri-Pacer Press
P.O. Box 840111
Pembroke Pines, FL 33084-2111
Telephone: 954-432-9683

TABLE OF CONTENTS

FOREWORD

In the past ten years, more than one hundred would-be authors of aviation books have asked me to help them get their books published. The range of the work has been remarkable, from the biographies of astronauts to novels to lovingly compiled histories of obscure aircraft. Almost all of the authors had one thing in common: they did not have the least idea how to begin. Most had experienced rejection in one form or another, from agents or publishers, and most were willing to go to any lengths to get published.

How I wish I could have had copies of Chevy Alden's *How To Write, Publish & Sell Your Own Aviation Books* to send them! I did the best I could, and even prepared a little seven page outline of suggestions. It seemed to have worked with about twenty percent of those who used it. If I had been able to give them Chevy's book instead, they all could have been successful.

The world of aviation is filled with thousands of interesting stories, and more are happening every day. There are enough ideas to fill whole libraries with aviation fiction and nonfiction, and there is a huge market out there for the material. Pilots like to read about flying.

The average person's difficulty in getting published lies in the arcane world of editors, publishers and agents, who all prefer to turn to already published authors. Many of them do not have the background to recognize good aviation material when they see it in manuscript form. The solution is to build a reputation for yourself by publishing yourself. When you have a self-published book or two under your belt, along with some book reviews that will come your way, you'll have something to attract an editor's eye. You will no longer be an "unpublished author" who deserves automatic refusal. You will have shown that you have the intelligence and the perseverance to create a book. A manuscript accompanied by a previously published book will get editors to read your work - and that is the essential first step.

In *How To Write, Publish & Sell Your Own Aviation Books*, Chevy Alden does not give you a hypothetical situation. Instead he shows you exactly how he achieved success. I was impressed with his attention to detail, and with the methodical approach to the problem. It is virtually a fool-proof approach, is relatively inexpensive, and keeps you out of the hands of the rip-off artists of the various vanity presses that prey on authors.

Besides the satisfaction of seeing your book in print, and in the leverage this gives you with publishers, you can actually make a profit self-publishing, for instead of the chancy royalty reports of ten percent of the sales price, you keep the entire difference between the sales price and the cost of making and selling the book. The mark-up can be significant.

We need more aviation books by more writers on more subjects. If you feel you have a book in you, Chevy Alden has prepared a road-map to get it into press. I recommend it highly.

Walter J. Boyne
Author/Historian
Former Director of the National Air and Space Museum

CHAPTER ONE

THE SECRET TO GUARANTEED PUBLICATION

The adventurous field of aviation continually supplies writers with enough good material to create thousands upon thousands of unique aeronautical books. Whether you want to write fiction or nonfiction, you can easily join the ranks of today's growing list of aviation authors by first learning the secret that leads to guaranteed publication. That's right! There is a trade secret within the publishing industry that can guarantee the publication of your own aviation books.

This tightly-held trade secret is not something new. In fact, Mark Twain used this method of publication for his American classic *The Adventures Of Huckleberry Finn.* Other authors who have published books using this method include such greats as Upton Sinclair, Zane Grey, Edgar Allan Poe, Henry David Thoreau and Rudyard Kipling.[1]

Before I discovered this trade secret, my first aviation book was rejected by eighty-four publishers. Within a year of my discovery, however, *Black Falcon* was successfully published in a hardcover edition, received dozens of favorable book reviews in newspapers and magazines, and even found its way onto President Bush's bookshelf in the White House. *Black Falcon* is still in print today, years after its original publication date back in October 1988.

What is this trade secret? Read on! By the time you finish reading *How To Write, Publish & Sell Your Own Aviation Books* you will be privy to information that can change your life and put you on the road to becoming a published aviation author.

You might be asking yourself, "Why do I have to learn some trade secret in order to get published? Why can't I just send my manuscript to one of the major New York publishing houses? After all, I know my book is good."

Well, your book just might be good - just like I always knew *Black Falcon* was good. But that doesn't mean anything to the New York publishing industry - not unless you are a celebrity or you already have a publishing track record. Unfortunately, in today's complex world of multinational publishing superconglomerates, this is the way it has to be. Mainstream publishing no longer considers writing to be an artistic endeavor, but rather a business - a very big business.

Today's major publishing houses have stockholders to please, and stockholders are pleased by profits. In most instances, profits cannot be made from books written by unknown authors. With well over 50,000 new titles being published each year, and with most of them losing money, it is no wonder that major publishing houses are cautious about which authors they sign book contracts with. They want to cut their losses right from the start by mainly signing on authors who have publishing track records. Rarely are unpublished authors taken seriously. It's a case of the infamous *Catch-22* of the publishing world: "If you don't have an agent, you can't get published. If you aren't already published, you can't get an agent."

This is just one of the many roadblocks preventing unpublished authors from achieving their goals of becoming published authors. The proper terminology for this situation is called *lack of a publishing track record.* I will be mentioning track records throughout this text because track records are what separate amateur writers from professionals. And any writer who has even the slightest desire of getting published is really saying that he or she wants to be a professional writer, even if only on a part-time basis. If not, then there's no reason to waste the time and effort needed to get published.

So how does an unpublished author go about getting a track record if the major publishers won't take a gamble on him or her? The answer lies within this chapter - but first let me reassure you that even bestselling authors such as Tom Clancy, master of the military techno-thriller genre, had to face this problem before getting published for the first time. In Clancy's case, *The Hunt For Red October* was rejected by every major New York publishing house it was sent to. Why? Because Clancy lacked a publishing track record. He hadn't yet proven that his writing would sell.

Did Clancy give up? Not in the least. Instead, he did research and found a small segment of the publishing industry that *does* consider unpublished authors - the small press industry, also known as independent publishing.[2]

According to *Small Press* magazine, there are several thousand independent publishers, compared to only about a hundred major publishers.[3] Yet, even with all these independents operating in the United States, they only account for a small fragment of the 50,000 new books that are published each year. The good news is that they are willing to take a chance on unpublished authors. In fact, this is their mainstay for finding new material to publish. Why? Because most small presses are family-owned enterprises. They love the glamour and the prestige associated with literary endeavors, and they consider books to be artistic creations - not just products to sell to consumers like meat and potatoes.

Tom Clancy found just such a publisher in his own hometown of Annapolis, Maryland, when he came across Naval Institute Press. Prior to publishing *The Hunt For Red October*, Naval Institute Press had never before published a novel, sticking strictly to nonfiction titles pertaining to naval warfare, nautical-related technical manuals, books about naval weaponry and ships, biographies about historic naval personalities, and histories of the United States Navy. When they read Clancy's manuscript, however, they saw something that the major publishers had missed - a new style of writing that has since become known as the military techno-thriller genre. They took a gamble on Clancy and bought the rights to his book - a first for them, as well as a first for Clancy.[4]

Publishing history was made when *The Hunt For Red October* made the bestseller list. Clancy gained celebrity status overnight and jumped the fence to the major New York publishers.[5] He had made it to the big leagues by first proving his worth in the minors. He had built himself a track record - and an impressive one at that.

As you can see, submitting a manuscript to a small press gives an unpublished author his or her best chance of getting published. But it is still just that - a chance. There is no guarantee.

And now for the secret of how you can guarantee the publication of *your* books! An even smaller segment of the publishing industry exists that is a mere corner of the independent publishing field. This corner is known as self-publishing and it is the tightly-held trade secret that I've been referring to - the secret that leads to guaranteed publication.

Simply stated, self-publishing involves an author setting up his or her own independent publishing company at home (which is what I did with Tri-Pacer Press in 1988) and contracting out the printing and binding of the books to specialized book printers. Through a strategic marketing plan, the books are then sold for a profit - the author's profit! Self-publishing is easy, fun and a great way to learn about the publishing industry from an insider's point-of-view. It's also a fantastic way to build a track record. Mark Twain did it with *The Adventures Of Huckleberry Finn*, I've done it with *Black Falcon*, and several thousand other authors are doing it as well.

Almost anybody who has the concentration to write a book can self-publish a book. In fact, it's easier to self-publish a book than it is to write a book in the first place. I know this to be true because I've done both - *several times*. You don't have to be a genius to self-publish a book, nor do you need an extensive business background to run a small publishing company. Take me, for example. I'm a blue-collar worker who spends five days a week servicing corporate jets at Fort Lauderdale-Hollywood International Airport. For me, self-publishing has always been a part-time venture done in my spare time.

You'd be amazed at what can be accomplished in one's spare time. Figures 1-1 and 1-2 show the publishing track record I built up by the end of my first five years as a self-publisher (which covered the period from the end of 1988 to the beginning of 1994). Pretty impressive, wouldn't you say? Prior to acquiring that track record, I was unpublished and unknown. Since that time, however, I have become a lecturer and award-winning author. My second book, *How To Get Published - Guaranteed*, won the National Association of Independent Publishers' 1994 Book Awards competition and also received a Certificate of Merit for Outstanding Accomplishment from *Writer's Digest* magazine when I entered *How To Get Published - Guaranteed* in their 1994 National Self-Publishing Awards competition.

I kid you not when I tell you that self-publishing is a viable way to publish your books. This small segment of the publishing industry is growing all the time, especially since the advent of the personal computer - although a computer is not really necessary to self-publish a book (and I'm a prime example of this, since I didn't learn how to use a computer until after my second book was finished).

Self-publishing is becoming so popular that feature arti-

AUTHOR FACT SHEET

CHEVY ALDEN lives in Broward County, Florida — DOB: May 22, 1958

BIOGRAPHY: Aircraft Refueler, High School Substitute Teacher, Novelist, Aviation Journalist, Pilot, Sailor, Adventurer and Self-Styled Thrill-Seeker.

PUBLISHED WORKS:

Down's Syndrome In The 1980s – Expect The Most, published in *Centerpoint* (August 1988), reprinted in *The Advocate* (January 1989), reprinted in *The Parent Press* (November/December 1989). A touching article about the author's daughter, who was born with Down's Syndrome.

Black Falcon, published in a hardcover edition by *Tri-Pacer Press* (October 1988), reprinted in hardcover by *Tri-Pacer Press* (March 1991), scheduled reprinting in hardcover by *Tri-Pacer Press* (Fall 1994). A shocking and suspenseful look into South Florida's deadly smuggling trade. This action-adventure novel has received much acclaim from book reviewers. A fast-paced thriller.

Untitled Editorial, published in *US Aviator* (April 1991). The author's viewpoint about the personal value of sport flying and the need to make financial sacrifices in order to enjoy this expensive hobby. The author is an active sport pilot.

Pride, Professionalism The New Standards For The FBO Front Line, published in *Aviation International News* (November/December 1992). A full-length article that looks at the current status of corporate aviation from a rampside perspective. The author works as an aircraft refueler within the corporate aviation industry.

Untitled Editorial, published in *Aviation International News* (January/February 1993). A comparison of wages and benefits paid to aircraft refuelers who work in large cities as opposed to those who work in small towns. The author works as an aircraft refueler within the corporate aviation industry.

Untitled Editorial, published in *US Aviator* (January 1994). A report about the cancellation of the 1993 Miami Air Show.

How To Get Published – Guaranteed, scheduled for publication in a trade paperback edition by *Tri-Pacer Press* (Fall 1994). A self-help manual that tells authors how to assure the publication of their own books. Based on the author's personal experiences while seeking publication for *Black Falcon.* Topics include writing, editing, publishing and much more. Enriched with more than 100 photos, figures, illustrations, documents, publishing forms and sample letters.

WORKS IN PROGRESS:

Antique Wings In Modern Skies. A full-length article about The Aerial Sign Company, a banner-towing operation that utilizes the world's largest fleet of working-class antique aircraft.

Death Wreath. The first sequel to *Black Falcon.* This air-sea adventure novel revolves heavily around the Federal Witness Protection Program and the undercover pilots who fly smuggling missions for the U.S. Drug Enforcement Administration.

FUTURE WORKS:

Antique And Classic Aircraft Across America. A book about ordinary people who restore and fly antique and classic aircraft as a hobby.

Avenge. The second sequel to *Black Falcon.* Like the rest of the series, this book is an air-sea adventure novel that revolves heavily around flying and sailing in South Florida.

Figure 1-1

PAGE 2

How To Promote & Market Self-Published Books. A self-help manual that tells authors how to publicize and sell their own books. Based on the author's personal experiences while promoting and marketing *Black Falcon.* Topics include book reviews, book shows, guest appearances, press releases, special events and much more. Enriched with documentation of actual case studies.

How To Write, Publish & Sell Your Own Aviation Books. A combined edition of *How To Get Published – Guaranteed* and *How To Promote & Market Self-Published Books* that is geared specifically towards aviation-oriented books.

How To Write, Publish & Sell Your Own Nautical Books. A combined edition of *How To Get Published – Guaranteed* and *How To Promote & Market Self-Published Books* that is geared specifically towards nautical-oriented books.

Mountains In The Sky: True Tales Of A Sport Pilot. A book-length collection of aerial adventures experienced by the author.

Path To The Left-Seat. A multi-book series dealing with various piloting jobs.

Unplugged: An Autobiographical Odyssey Through One Author's Life. The autobiography of Chevy Alden – author, pilot, sailor, thrill-seeker and worldwide adventurer.

GUEST APPEARANCES: Guest Speaker during *National Book Week* (1988) at the Broward County South Regional/BCC Library. Sponsored by The Friends Of The SR/BCC Library.

Guest Speaker during a monthly meeting of *The Backroom Writers* (October 1990) at the Palm Beach Community College Library Learning Resource Center. Sponsored by The Backroom Writers.

Guest Panelist during the *1st Annual BookFest Of The Palm Beaches* (1991) at the South Florida Fairgrounds Convention Center. Sponsored by the Palm Beach County Library Association and the Palm Beach County Literary Coalition.

Guest Speaker during *National Library Week* (1992) at the Harold C. Manor Library. Sponsored by the Palm Beach Community College Library.

Guest Speaker during the *18th Annual Sun 'n Fun EAA Fly-In* (1992) at Lakeland-Linder Regional Airport. Sponsored by the Experimental Aircraft Association.

Guest Speaker during the *19th Annual Sun 'n Fun EAA Fly-In* (1993) at Lakeland-Linder Regional Airport. Sponsored by the Experimental Aircraft Association.

Guest Speaker during the *20th Annual Sun 'n Fun EAA Fly-In* (1994) at Lakeland-Linder Regional Airport. Forum entitled *How To Write And Publish Your Own Aviation Books.* Sponsored by the Experimental Aircraft Association.

LITERARY HONORS: Honored Guest during the *3rd Annual Palm Beach Literary Forum & Author's Luncheon* (1989) at the Admirals Cove Clubhouse. Sponsored by the Palm Beach Community College Library.

Received a handwritten letter from *President George Bush* (April 1990) stating that he had a copy of *Black Falcon* in The White House. He also wrote that as a former naval aviator, he was definitely going to read the air-sea adventure novel.

Figure 1-2

cles about this subject have appeared in the pages of *The New York Times* (*Do-It-Yourselfers Carve Out A Piece Of The Publishing Pie* and *New Chapter In Book Publishing: More Authors Are Publishing Themselves And Finding The Public Will Buy*), *The Palm Beach Post* (*More Authors Becoming Own Publishers*), *U.S. News & World Report* (*Small Has Never Looked So Beautiful*), and *The Fort Lauderdale News/Sun-Sentinel* (*Book Takes Flight: Pines Author Not Only Writes Air-Sea Adventure Novel, But Publishes It Too*). These are just five examples of the dozens of articles I've collected since starting Tri-Pacer Press in 1988. There was even a television news broadcast on *CNN* on the evening of May 11, 1992, which discussed self-publishing and the success of several self-published authors.

This home-grown industry has been around for hundreds of years and it's not going away any time soon. Why? Because self-publishing works! (Caution: Please do not confuse self-publishing with vanity publishing, which is a fraud in most instances. I will tell you more about vanity publishing in Chapter Three, explaining in full detail why you should steer clear of this method of getting your books into print.) As I stated, self-publishing works. Take me, for example.Without any prior publishing experience, I successfully self-published my first book when I was thirty years old, my second book when I was thirty-five years old, and my third book when I was thirty-nine years old. That's three self-published books in a nine year period! In between all of this I built up an impressive track record that included numerous speaking engagements, personal appearances, author luncheons, book shows, and dozens of other special events - all through self-publishing, self-promotion and self-marketing. And all done on a part-time basis!

Right now you are probably filled with many questions about self-publishing, asking yourself, "How do I get started with self-publishing? How much time does it take? How much does it cost? How much can I expect to earn? How do I sell my books once they are in print? How do I get book reviews? How do I arrange personal appearances and speaking engagements?"

These are all valid questions and I will answer each one of them before the end of this book. Rest assured, I had the same questions when I first found out about self-publishing. Rather than letting the questions drag me down, though, I forged ahead, bought a copy of *The Self-Publishing Manual* by Dan Poynter, and started Tri-Pacer Press. Poynter's book provided me with enough information to successfully put *Black Falcon* into print and to start selling it to pilots. Within two years the first printing was sold out and I had to go back to press for a second printing. A third printing of *Black Falcon* was released in 1996.

The reason I wrote *How To Write, Publish & Sell Your Own Aviation Books* was to show you how somebody with absolutely no prior knowledge or experience in the publishing industry can indeed successfully self-publish his or her own aviation books. Is *How To Write, Publish & Sell Your Own Aviation Books* in competition with *The Self-Publishing Manual*? Not in the least bit! You see, Poynter's book describes the detailed, technical side of the self-publishing industry. My book, on the other hand, describes the human side of the picture - the emotions, the joys, the frustrations, the successes, the failures, the do's and don'ts of putting a book into print for the first time. I let you witness the trials and tribulations of a novice self-publisher during the formative years of such an endeavor. You will see how I turned my raw manuscript into a highly-polished aviation novel. You will see how I found an artist, typesetter and printer, and how I coordinated the three of them to get *Black Falcon* into print. You will witness the transformation of an unpublished author into a successfully published one. Best of all, you will learn how you can do the same thing!

Rather than being in competition with *The Self-Publishing Manual*, my book is a compliment to Dan Poynter's book and both should be used side-by-side as desktop references by any author who is seriously thinking about self-publishing his or her own book.

Some skeptics might claim that I was lucky with the success of *Black Falcon*. Let me tell you that luck had nothing to do with it! If I was lucky, *Black Falcon* would have been picked up by one of the eighty-four publishers who rejected the book in the first place. No, I was definitely *not lucky*. The success I've had with *Black Falcon* can be attributed to two things - the strong belief I've had in my writing and an almost fanatical belief in Poynter's book, *The Self-Publishing Manual*.

As you read *How To Write, Publish & Sell Your Own Aviation Books*, I hope you will find encouragement and inspiration. After all, this book was written for *you*! Use this book as a source of hope - like a personal guidebook that will lead you toward guaranteed publication. Make notes in this book. Highlight important passages. Learn from my mistakes. Learn from my successes. Read the book a second and a third time, and refer to it often. Never take *no* as a final answer. Keep a positive attitude about what you are doing. And keep saying to yourself, "If Chevy Alden can do it, I can do it, too!"

By the way, let me introduce myself to you. Chevy Alden is my pseudonym - the name I use for my writing projects, personal appearances and speaking engagements. Richard Cook is my real name - the name I use as the owner and editor-in-chief of Tri-Pacer Press. Welcome aboard! Now prepare yourself to embark on a magical journey into the fascinating world of guaranteed publication!

CHAPTER TWO

THE REWARDS OF BEING A PUBLISHED AUTHOR

Many rewards are bestowed upon authors once they become published. For instance, published authors are considered to be experts on the subjects they write about. This, in turn, opens doors to other avenues of endeavor, such as freelance magazine writing, speaking engagements, guest appearances and consulting work. Additional rewards include such things as emotional satisfaction, the ability to touch people's lives, admiration from others and monetary rewards.

During my formative years as a published author, I was fortunate enough to have experienced every one of the above mentioned rewards. Self-publishing made it all possible. Before I actually get into the meat of this book, which explains in graphic, autobiographical detail how I self-published *Black Falcon* and how you can do the same with your aviation books, I would first like to tell you about some of the rewards that came my way because of self-publishing. I think you will find this chapter extremely inspirational - something to look forward to once you become a published aviation author yourself.

Probably the most cherished rewards for me are the times when total strangers tell me how much they've enjoyed reading *Black Falcon*. Their admiration creates a powerful and pleasurable high. For example, one day while I was at the fuel storage area at the airport transferring jet fuel from an underground storage tank into a fuel truck, the chief of the airport's Crash, Fire & Rescue squad came over to me to say that he had just read *Black Falcon* and that he really enjoyed it. That really brightened my day, especially since the comment had come totally out-of-the-blue. I didn't even know he had a copy of my book.

On another occasion, a pilot came up to me to say that his boss (the owner of a corporate King Air 200) had read *Black Falcon* while on a fishing trip in the Bahamas, and that his boss felt I was wasting my time and talent fueling airplanes for a living. His boss felt that I should be writing novels full-time. That was one of the nicest compliments I had ever received from anybody. The pilot then added that not only did his boss enjoy my novel, but that all of his fishing buddies had read *Black Falcon* during their week-long vacation at Chub Cay and that they all came to the conclusion that *Black Falcon* had *made their week*! I was stunned by that compliment - utterly floored. I didn't know what to say.

You see, I don't get this kind of praise from people in my routine, work-a-day world of fueling airplanes. Pilots taxi their planes to the ramp, I fuel the planes, then they taxi out for departure. It's a job that must be done in order for planes to fly, but there's no praise involved. I enjoy my job at the airport (and it's paid my bills for over twenty years), but it's also nice to be appreciated by others. My writing takes care of my need for appreciation.

Another reward I've received is the knowledge that my writing actually touches other people's lives in very personal ways. For some people this means a few hours of reading enjoyment during their leisure time - a thrilling, fictional trip to places they ordinarily would not have access to, such as the cockpit of a tramp airfreighter, a cruise aboard a sailboat in the Bahamas, or a fishing trip in the Everglades. For others, my writing provides useful information that they can use during the course of their lives. It all depends on which book of mine they are reading, since I write both fiction and nonfiction.

But these aren't the only ways that my writing affects people's lives. One of the most fascinating true stories regarding this phenomenon was told to me back in 1990, approximately two years after *Black Falcon* was first published. What is so amazing about this story is that it happened to someone I had never met and who lived all the way down in Venezuela (in South America).

The whole situation started when a Venezuelan corporate pilot on a trip to Florida purchased a copy of *Black Falcon*. He brought the book home to Venezuela, read it, then passed it around to other pilots who were based at the same airport as he was. One of those pilots happened to fly a helicopter, which meant that he did quite a bit of low-level flying. Because there's a lot of low-level flying within the story line of my book, all the other Venezuelan pilots started calling the helicopter pilot *Black Falcon* as a nickname. The nickname stuck and he has been known as *Black Falcon* ever since. Think about that for a moment. A fictional story that I created in my mind was the basis for somebody else's nickname - somebody who lived on another continent! It's mind boggling!

There are other types of rewards bestowed upon an author once he or she gets published - rewards that can help build a writing career into something special. Once of these rewards is the fact that a published author will usually be considered an expert in his or her chosen field of endeavor. This can lead to many new opportunities, which will then lead to even more opportunities.

For example, during the summer of 1992 I wrote a lengthy letter to the editor of *Aviation International News* (one of the world's leading aviation magazines) seeking publicity for my novel in the form of a book review. I explained in the letter how I wrote and self-published *Black Falcon*. Instead of getting publicity, I got an offer to write a couple of freelance articles for *Aviation International News*. I was more than thrilled at the opportunity!

Needless to say, I accepted the offer to write for *Aviation International News*. My first full-length article was published in the November/December 1992 issue, followed by a short editorial in the January/February 1993 issue.

The full-length article in *Aviation International News* ended up giving me credibility as an aviation journalist - not just an aviation novelist. This, in turn, lead to other opportunities. How? The credibility made it easier for me to get interviews with prominent aviation personalities and industry leaders, making it simpler to do research for future aviation-related magazine articles and nonfiction aviation books.

For example, while working on my second book I was also doing research for an article about The Aerial Sign Company, a banner-towing outfit operating out of North Perry Airport that utilizes the world's largest fleet of working-class Piper taildraggers (classic aircraft that date as far back as the 1940s). With a copy of *Black Falcon* in hand, as well as a copy of the November/December 1992 issue of *Aviation International News*, I was able to show my writing credentials to the owner of The Aerial Sign Company, who then agreed to do a two-and-a-half hour taped interview. He then gave me a personal tour of his maintenance hangar and storage facility, as well as the opportunity to view firsthand (and up-close) the actual banner-towing operation going on at North Perry Airport. I also got the chance to fly one of the planes - a Piper J-3 Cub. Figure 2-1 shows a photo of that plane. *(Did I mention that I'm an avid sport pilot and that I love flying airplanes more than just about anything else?)*

On another occasion I had the opportunity to do re-

Figure 2-1

search for an article about Florida Air Cargo, an airfreight operation based out of Opa Locka Airport that utilizes a pair of Douglas DC-3s to haul cargo to and from the Bahamas. I rode in the cockpit jumpseat during a flight from Opa Locka Airport to Freeport International Airport with a cargo of baby chickens. Then, while the plane was being off-loaded on the ground in Freeport, I interviewed the captain and copilot to get information for the article. I again rode in the jumpseat during a flight from Freeport International Airport to Fort Lauderdale-Hollywood International Airport with a load of fresh fish, followed by a repositioning flight back to Opa Locka Airport. Figure 2-2 shows a photo looking through the DC-3s windshield while taxiing onto Runway 9L at Opa-locka Airport just prior to our 7:00am takeoff. Figure 2-3 shows a photo of the DC-3 on the ramp at Freeport International Airport while waiting for the cargo to be off-loaded. *(Do I enjoy the benefits of being an aviation journalist? You bet I do!)*

Being a published author also puts you in touch with

Figure 2-2

Figure 2-3

other published authors. For example, my first introduction to Henry M. Holden was over the telephone when he called to order a copy of *Black Falcon* after reading a book review about it. Holden had had a series of three books published by a major publishing house, but the publisher let the entire series go out of print after a very short time. When I told Holden about how I self-published *Black Falcon,* he got very excited about starting his own aviation publishing company. I finally got to meet Holden several years later at the Sun 'n Fun EAA Fly-In, where he was selling autographed copies of his books about female aviators. He had self-published the books through his own company, Black Hawk Publishing, and was doing quite well with them. Holden said that he was working on a very detailed book about the history of the Douglas DC-3 and that several hundred photos were going to be included in the text. He also said that he was looking for a larger publisher to take on the project - someone who would publish the book on a grand scale. Well, *The Legacy Of The DC-3* rolled off the press in 1997 and it was definitely done up in grand fashion, including a foreword by Donald W. Douglas, Jr. (son of the founder of the Douglas Aircraft Company). Published by Wind Canyon Publishing, the book also included about a dozen DC-3 photos that I had shot at various South Florida airports. I thought that was pretty cool, especially since the DC-3 has *always* been my favorite aircraft. And it all came about because I was a published aviation author.

Being an expert in a certain field of endeavor allows published authors to arrange speaking engagements and personal appearances. For example, I've appeared before groups of pilots, sailors and writers - sometimes giving speeches, other times just answering questions and autographing books. Let me tell you, it feels almost criminal to get paid for talking about stuff that I love talking about - namely airplanes, sailboats and books! But, hey, it's all part of the authoring game, so I happily go along with it and take what opportunities come my way. To top it all off, I usually sell quite a few books after my talks, which only serves to add gravy to the meat. *(Can you imagine having so much fun and getting paid for doing it? It's great!)*

Let me go off on a tangent here for just a few moments to tell you about an exciting opportunity that came my way because of one of my aviation-related speaking engagements. First of all, let me explain briefly how I chose the name of my publishing company. During the 1950s, the Piper Aircraft Corporation built a series of single-engine airplanes called Tri-Pacers. I've always had a strong affection for these classic, tube-and-fabric four-seaters, and for a long time I thought that someday I'd be able to buy one. By calling my company Tri-Pacer Press, I've been able to keep my dream alive and in the forefront on my mind (although I might decide on purchasing another type of airplane once I'm financially able to become an aircraft owner).

What you have to know, though, is that up until April 20, 1993, I had never had the opportunity to fly a Tri-Pacer. That situation changed on the morning of April 20th. You see, that's the morning I was walking among the hundreds of classic airplanes at the Sun 'n Fun EAA Fly-In in Lakeland, Florida (an air show to which I had been asked to be a guest speaker). While walking among all the restored airplanes, I met the owner of a white and red Tri-Pacer who had flown his four-seater all the way down from Ohio just to attend the week-long event. After talking with him about his background as an aircraft mechanic and sport pilot, and my background as an aviation novelist and sport pilot, he invited me for a flight aboard his Tri-Pacer the following morning. I was elated!

My presentation went well that night, but my mind was on the next day's flight. When the sun finally came up, I met the owner of the Tri-Pacer at the tie-down area. After a thorough preflight, we both climbed into the restored Piper. Figure 2-4 shows a photo of us just before we taxied out of the parking spot. We then proceeded to the active runway at Lakeland-Linder Regional Airport.

When the Tri-Pacer was lined up on the runway's centerline, the owner turned to me and said, "It's all yours." Up until that moment, I had no idea he was actually going to let me fly his airplane. I just figured he was going to take me for a short flight around the traffic pattern. But *fly* I did! I took off on an easterly heading, then turned south, where I felt out the controls by doing shallow turns, steep turns, climbs, descents, stalls, and even a short cross-country flight to Bartow Municipal Airport, where I did a series of touch-and-go landings before heading back to Lakeland-Linder Regional Airport.

I had finally flown a Tri-Pacer! Figure 2-5 shows a photo of me sitting in the cockpit of that very special airplane. I can't say for certain that the owner of the Tri-Pacer trusted me with his aircraft because I was a published aviation author. Perhaps it helped, perhaps it didn't. What matters is that I had given him enough of an impression of knowing about airplanes (and about flying) for him to have given me the go-ahead to fly his aircraft. The opportunity arose and I took it.

One reward which *does* come from being a published author is consulting work. For example, I have had a few minor consulting jobs in the field of self-publishing, whereby I've guided other authors who were setting up their own self-publishing ventures. I haven't done too much of this in the past and I don't plan on doing too much of it in the future, but occasionally I'll accept a consulting job if the situation is right and I'm not too busy with other projects.

Another reward bestowed upon an author once he or she is published is the ability to reach important people via his or her writing. For instance, I reached President George

Figure 2-4

Figure 2-5

Bush and actor John Travolta through *Black Falcon*. Not only did I reach these very busy people, but I got personal responses as well.

Figure 2-6 shows the envelope I received from the White House on April 18, 1990. Inside the envelope was a handwritten letter from President Bush, which stated, "The V.P. gave me *Black Falcon* complete with that very nice note from you. Yes, as a former pilot, I'll not only read it, but I know I'll enjoy it. My sincere thanks - best wishes." Figure 2-7 shows a copy of this letter. The original letter has been valued at close to $1,000, but I will never sell it to autograph collectors because it holds too much emotional value.

Figure 2-8 shows a letter I received from the personal assistant to John Travolta on November 28, 1989. In the letter, she stated, "Thank you very much for sending *Black Falcon*. Mr. Travolta read it and enjoyed it. Commitments, personal projects, and a heavy schedule do not permit his involvement despite the indeed fun prospect of flying a Con-

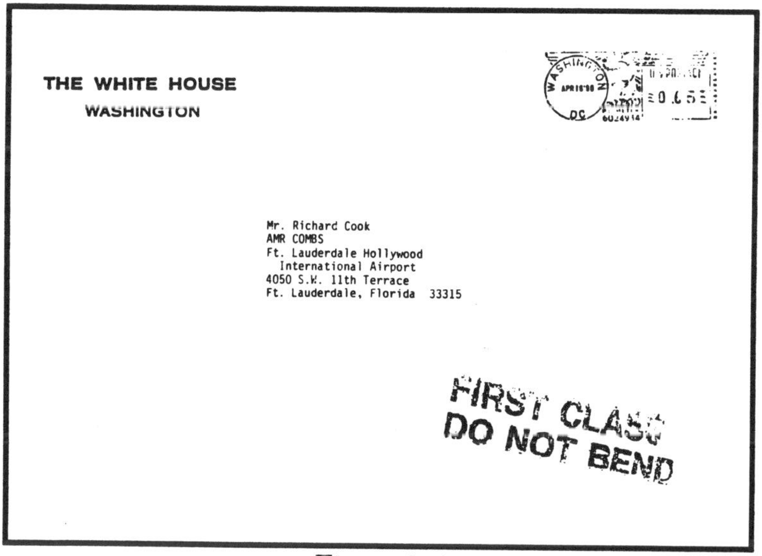

Figure 2-6

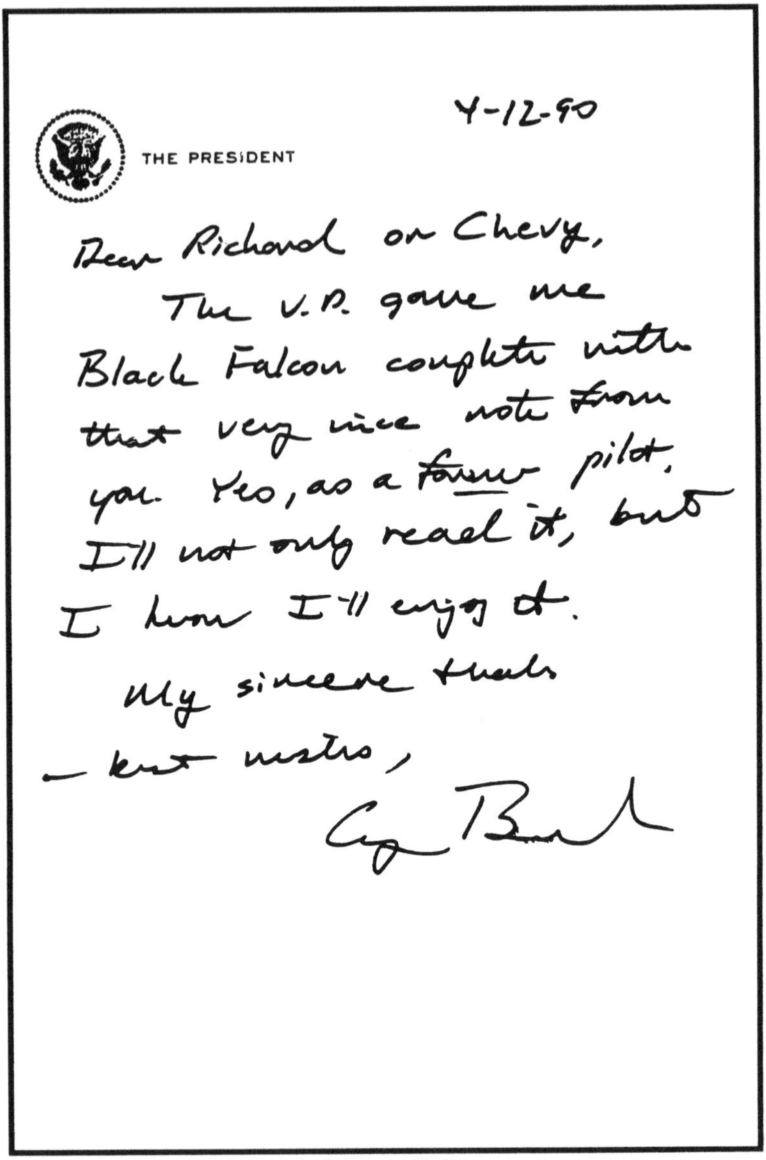

4-12-90

THE PRESIDENT

Dear Richard or Chevy,

The V.P. gave me Black Falcon complete with that very nice note from you. Yes, as a former pilot, I'll not only read it, but I know I'll enjoy it.

My sincere thanks

– best wishes,

Figure 2-7

John Travolta

November 26, 1989

Dear Mr. Cook,

Thank you very much for sending Black Falcon. Mr. Travolta read it and enjoyed it.

Commitments, personal projects, and a heavy schedule do not permit his involvement despite the indeed fun prospect of flying a Constellation.

Thank you for your interest and frankness.

Sincerely yours,

Joan Edwards

Joan Edwards
Personal assistant to
John Travolta

Figure 2-8

stellation. Thank you for your interest and frankness."

This response deserves an explanation in order for you to understand it. During the early 1980s, I had read an article about John Travolta that said he was an avid pilot, as well as being a famous movie and television actor. Like myself, Travolta had a strong affection for older aircraft, especially radial-engined airliners from the 1940s and 1950s. In fact, he had once purchased a run-down Lockheed Constellation (the queen of the skies during the piston era) for the purpose of restoring it and flying it around the world, but that project was never completed.

Well, when I finished writing *Black Falcon* (a novel that revolved around radial-engined airfreighters), I thought about Travolta's dream of flying a Constellation. I thought to myself, "Wouldn't it be neat if *Black Falcon* could be turned into a movie and the main character (Rob Jensen, an airfreighter pilot) could be played by John Travolta? And wouldn't it be neat if instead of using a Douglas DC-4 as the primary airfreighter (like it is in the novel version of *Black Falcon*), a Lockheed Constellation could be used in the film version?" After all, both airplanes were four-engined transports from the same era. That way, Travolta would finally get to fulfill his dream of flying such a plane.

I decided to contact John Travolta and explain the idea to him. After spending a lot of time tracking down his home address, I sent him a proposal for a movie deal. The letter shown in Figure 2-8 was the result of my efforts. In the long run, I failed at selling Travolta the movie rights, but at least I tried.

Another reward bestowed upon a published author is the fact that he or she will finally have enough clout to be taken seriously by the New York publishing industry. A good case in point is my dealing with Random House during the spring and summer of 1990 - almost four years after I had originally been rejected by eighty-four publishers (including Random House) and two-and-a-half years after I had self-published *Black Falcon*. I had decided at that time to try to sell the rights to my novel to one of the major New York publishing houses to see if I could go mainstream. After carefully choosing specific editors at fifteen major companies, I sent each editor the following letter:

"What book is President George Bush reading these days? As a former naval aviator, he is enjoying my air-sea adventure novel *Black Falcon*, a thrilling tale revolving around South Florida's deadly, aerial drug trade."

"*Black Falcon* was published by Tri-Pacer Press, an independent publishing company founded in 1987 to market authentically-written aviation novels to Florida pilots. Besides being the author of *Black Falcon*, I am also the editor-in-chief of this publishing firm."

"Book reviews on *Black Falcon* have been very good and the first printing is nearly sold out. A second printing is in the planning stages and should be available to pilots at Florida-based flight shops and aviation bookstores by late summer."

"Now that the President of the United States is reading *Black Falcon*, the time is right to market this book nationally to the general public via standard bookseller distribution channels. Tri-Pacer Press is not equipped to handle such a wide-scale distribution operation; therefore, we are selling the rights to *Black Falcon* to the highest bidder. I am offering you the opportunity to bid on these rights."

"Please let me know if you are interested in this proposal. A copy of *Black Falcon* can be sent upon your request. I look forward to receiving your reply."

What I was trying to do was hold an auction for my book, which is how agents sometimes sell manuscripts. I figured I had nothing to lose by trying. Well, guess what? I wasn't brushed off as a phony. In fact, I received a handwritten reply (very rare) from Random House, which stated,

"We'd be glad to consider *Black Falcon*. Please send us a copy. How many copies have you sold, and how many are going into print?" (You can see a copy of this letter within the pages of *How To Promote & Market Your Own Books*, which is the sequel to the book you are now reading.) Even though *Black Falcon* was eventually rejected by Random House, I felt good that I was finally able to make a small breakthrough into the world of big-time book publishing.

Another reward that comes from being a published author is the opportunity to meet all kinds of interesting people. During my first five years as a published aviation author I got the chance to meet military pilots, airline pilots, bush pilots, sport pilots, aircraft mechanics, aeronautical engineers, law enforcement officers, secret service agents (which is how I managed to get *Black Falcon* into the White House), as well as other published aviation authors. One common thread that tied all these people together was the fact that each one of them had something important to say - something that should be written down for others to read. A second common thread was that they all enjoyed reading aviation books.

Which brings me to the final reward - the fact that people will pay hard-earned money to a published author to purchase his or her latest book, as well as for his or her presence at special events. Figure 2-9 shows the first check I ever received for *Black Falcon*, which is dated March 31, 1988 - seven months prior to the publication of the book. How did I get this check before *Black Falcon* was even in print? I had made an advanced mock-up of the cover art and had gotten permission to put it on display at Fort Lauderdale-Hollywood International Airport inside the Executive Terminal building of the FBO where I worked, along with excerpts from *Black Falcon* and advanced order forms. On March 31, 1988, a pilot from Longwood, Florida, walked inside to pay for his fuel, saw my display, read the excerpts,

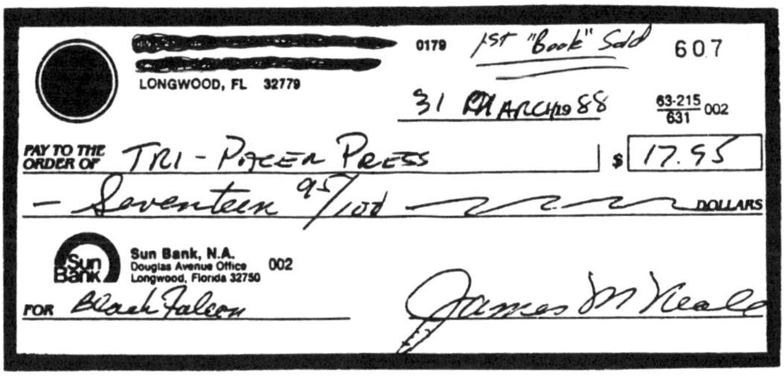

Figure 2-9

filled out an order form and sent a check. The pilot was a total stranger to me, yet he paid money up front for a copy of my novel. I knew that I was onto something exciting, and I was right - the checks have never stopped coming in!

Figure 2-10 shows the first check I ever received just for speaking. In this particular case, the check was sent to me by the Palm Beach County Library Association, the primary sponsor of the annual BookFest of the Palm Beaches. I had been one of the authors who participated in an hour-long *How To Write And Publish A Book* panel discussion at the 1991 event. This involved answering questions from an audience of approximately two hundred people. Easy stuff! A check in the amount of $100 arrived in the mail a couple of weeks later - a total surprise to me since I hadn't been told that I was getting paid to be a panelist. *(Hey, that's the kind of surprise I like!)*

By now, you should realize that self-publishing is a viable, worthwhile, and acceptable way to get published - especially for an author who is just launching his or her

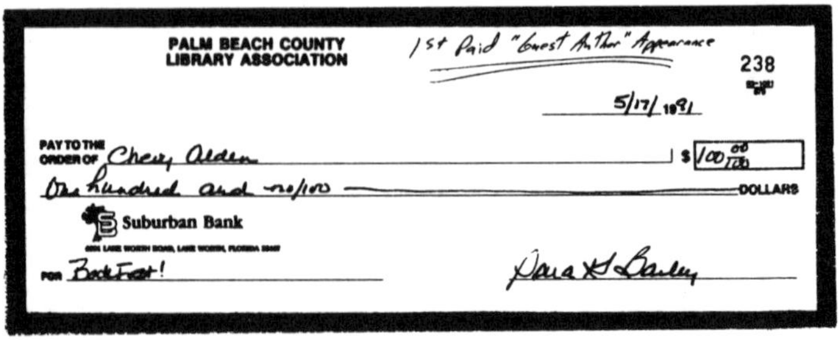
PALM BEACH COUNTY
LIBRARY ASSOCIATION

1st Paid "Guest Author" Appearance

238

5/17/1991

PAY TO THE ORDER OF Cheryl Alden $ 100 00/100

One hundred and no/100 DOLLARS

Suburban Bank

FOR Book Fest!

Figure 2-10

writing career.

The rest of this book will show you the meat of the self-publishing process - the very stuff that allowed me to receive all of the rewards you just read about. So without taking up any more of your time, please turn to the next chapter.

CHAPTER THREE

ROUTES TO GETTING PUBLISHED

There are four routes a writer can take in order to get published. These routes are major publishing, independent publishing, vanity publishing and self-publishing. Although I am biased towards self-publishing, I will give each option equal play within this chapter, which should help you decide which route is best for you.

Major Publishing. The major publishing houses are the ones most authors think about when it comes time to submitting their manuscripts for publication. Most of these companies (about one hundred of them in all) are located in the New York area, but a few are scattered around other parts of the country as well. These are the superconglomerates (*nice word, huh?*) of the publishing industry - the major leagues.

Most authors dream about writing a bestseller and having it published by one of the major publishing houses, thereby becoming instantly wealthy. I had that dream with *Black Falcon*. Tom Clancy had that dream with *The Hunt For Red October*. Although both Clancy and I failed in our attempts at attracting major publishers with our respective first manuscripts, it doesn't mean that all first-time authors will be rejected outright by these companies. That is the good news. The bad news is that *most* first-timers *will* be rejected outright by the majors.

If you choose to go this route in seeking publication, you will have a tough time trying to gain their attention without having either a publishing track record or an agent. Be prepared to equip yourself with the knowledge of how to play their game according to their rules. Many excellent books have been written about these rules. A sampling of titles includes *The Writer's Survival Manual*, *The Business Of Being A Writer*, *How To Be Your Own Literary Agent* and *How To Understand & Negotiate A Book Contract*.

Why do the major publishers force authors to play by their rules? Economics! Major publishers are in business for one reason - to make as much money as possible. These companies are no longer owned by individuals or families (as they once were), but rather they are owned and controlled by stockholders who demand high returns on their investments. They want books that sell well - books that guarantee profits.

This is why so few unpublished authors are able to break into the ranks of big-time publishing. The simple economic truth is that books written by first-timers rarely break-even when published by major companies. On the other hand, books by well-known authors (such as Ludlam, King and Steel) almost always earn profits - usually in the millions. Naturally, if an editor at a major publishing company has a choice, he or she will pick the proven author over the unproven author every time. The stockholders demand this.

If you still plan on seeking publication from a major publisher despite the gloomy outlook, prepare yourself to accept a lot of rejection. Why? Because if you are not prepared for the rejection you will surely receive, it will feel like a sharp knife piercing through your heart. You see, in this industry, even well-known authors have experienced rejection at some point during their writing careers. It's a fact of life for authors.

An article written by Russell Galen for *Writer's Digest* magazine explained that during his career as a New York literary agent, he had read many manuscripts that he felt were potentially good books, but for some reason they just didn't meet the criteria set up by the majors. Galen went on to explain that in many cases these same manuscripts were either picked up by independent publishers or self-published, and that many did very well after coming out in print.[6]

The Hunt For Red October and *Black Falcon* are two perfect examples of this point. Clancy's book became a bestseller after being published by an independent publisher, while my book did very well as a self-published book. Another example is *Flight Of The Intruder*, a fast-paced first novel by Stephen Coonts. This aviation-oriented thriller was rejected by thirty-four major publishers before it was picked up by Naval Institute Press - the same small press that published Tom Clancy's first novel. *Flight Of The Intruder* went on to sell over a million copies and later became a major motion picture. Book and movie royalties allowed Coonts to buy an old Stearman biplane and fly it around the country on a three-month aerial adventure, which he then wrote about in his first nonfiction book, *The Cannibal Queen*.[7]

The book industry is abound with interesting stories about major works that were rejected time after time. As reported by Debbie Sontag in *The Miami Herald*, there was an experiment carried out by a Florida-based magazine whereby the magazine tried to prove the difficulty of getting published. They submitted a synopsis and the first three chapters of Marjorie Kinnan Rawlings' *The Yearling* to twenty-two publishing houses, changing the name of the manuscript to *A Young Cracker Comes Of Age*. This Pulitzer-Prize winning novel was rejected outright by thirteen publishers, including the book's original publisher. Only one publishing company even recognized the manuscript as *The Yearling*. Most just sent form letter rejections saying that they only accepted submissions sent in by literary agents.[8]

Although submitting an unsolicited manuscript to a major publishing house is a long-shot, every unpublished author should at least give it a try for the experience. Before trying, however, make sure you know who you will be dealing with at the publishing company you are submitting to. Does this sound confusing? It should, because it is.

Let's take Random House, Inc. as an example. Random House, Inc. is a subsidiary of Advance Publications. Random House, Inc. is then broken down into Random House Trade Division, Villard Books, Times Books and Turtle Bay Books - the last three of which are imprints of the Trade Division. Confused? We're not finished yet. Besides being a subsidiary of Advance Publications, Random House, Inc. also has subsidiaries of its own, including Alfred A. Knopf, Pantheon Books, Vintage Books, Crown Publishers, Clarkson Potter, Harmony Books, Orion Books, Random House Reference, Fodor's Travel Publications, Ballantine/Del Rey/Fawcett/Ivy Books, plus the children's division at many of these subsidiaries.[9] In other words, when you are submitting a manuscript to Random House, Inc. (or to any of the superconglomerates, for that matter), who do you send it to?

To find the answer, it takes a lot of research. It also takes a willingness to learn the rules of the game, because that's exactly what the submission process is - a *game*. Let's say that you've learned the rules well and submitted your manuscript to some of the major publishing companies. What are the odds that your manuscript will be published by one of them? *Writer's Market* quoted one superconglomerate as saying that 18% of their titles were written by previously unpublished authors. Most of those authors, however, were either already known within their specific fields or they had been in the news. The quote went on to say that authors should be agented prior to sending in their manuscripts.[10]

The fact remains that the odds of a major publishing house buying the rights to an unsolicited manuscript from an unknown, unagented and unpublished author are extremely lousy - but it does happen occasionally. This is why every unpublished author should at least give it a try, if only for the experience of dealing with the major publishers.

But remember this warning before you send in your manuscript: *Submission rules for manuscripts are very rigid. Major publishers will not bend the rules for you. You must bend for them.* If you do not follow the rules precisely, you will not only be wasting your valuable time and money, but you will be putting yourself on bad terms with the publishers right from the start.

Major publishers want professionally submitted manuscripts. Any manuscript not arriving in a professional manner will either be sent back to the author (if return postage was enclosed by the author) or it will be thrown out. You don't want either of these things to happen, especially when you consider the high cost of sending out manuscript submissions (high quality photocopies, envelopes, postage, return postage, etc.). I personally spent several hundred dollars while submitting *Black Falcon* to publishing houses back in 1986 and 1987.

To help save you some time and money, I have compiled an abbreviated list of ten rules for submitting manuscripts. Keep in mind that these are only the basic rules. Entire books have been written about this subject. I suggest you find one of these books and read it thoroughly if you choose to go this route for seeking publication.

Rule One: Buy the most current edition of *Writer's Market*. This book is updated annually because of the constantly changing status of the publishing industry. This valuable reference source contains the names and addresses of all the major publishing companies, plus the names of editors, the types of books each company is looking for, the average advance paid to authors, and the average royalty rate paid to authors.

Rule Two: Type your manuscript on white, 20lb. bond paper, double-spaced with wide margins. Cleanly correct all typographical errors. Type your name, the title of the manuscript and the page number at the top of each page.

Rule Three: Write a brief synopsis of your book. Type the synopsis exactly like the manuscript.

Rule Four: Make several high-quality photocopies of your manuscript and synopsis. If you are not sending out the entire manuscript, then you will only need to photocopy the first three chapters and the synopsis. Store the original manuscript and synopsis in a safe place. If you are using a computer, be sure to print the copies using a high-quality ink-jet or laser printer. Dot-matrix copies have never been acceptable.

Rule Five: Choose at least a dozen publishing houses that publish the same type of book you have written. These publishers will be the target of your first round of manuscript submissions. Some people in the industry advise against multiple submissions, but agents do it all the time, so you should, too. If you tried sending your manuscript to only one publisher at a time, it would take you a lifetime to submit just one book, especially since many publishers can take anywhere from three months to a year before responding to your submission. By using the multiple submission method, you can cover a lot of ground in a short amount of time. For instance, with *Black Falcon* I was able to cover eighty-four publishers in about a year's time.

Rule Six: Pick a specific editor at each publishing house you have chosen. Type an original query letter (a letter of submission) to each editor. In the query letter you should explain the type of manuscript you have written, the length of the manuscript, your credentials for having written it, why your book is different from all the other books in the same category, why your book is marketable, and a list of previously published works (including magazine and newspaper articles). Keep your letter short and simple. If it is more than one page, it is too long. Do not brag about your writing ability. Do not say that your book is sure to be a bestseller. Do not beg. But *do* write with confidence.[11] Make photocopies of each query letter and save them in a file folder. Figure 3-1 shows the query letter I used for *Black Falcon*. Notice that I was using my real name at that time and that the title of my manuscript was originally just *Falcon*. I will explain why I changed both later on in *How To Write, Publish & Sell Your Own Aviation Books*.

Rule Seven: Do not use staples or bindings of any type on your manuscript. Do no punch holes in your manuscript. Do not fold your manuscript. Put together a package containing the query letter, the synopsis and the first three chapters of the manuscript. Place this package inside a tight-fitting, shipping envelope. Be sure to include a self-addressed-and-stamped shipping envelope for the return of your manuscript package. If by chance you decide to send out your entire manuscript, place it inside a stationery box, then put the box inside a tight-fitting, shipping envelope.

Rule Eight: Send your shipping envelope to the publishing company. Be sure the envelope is addressed to a specific editor, otherwise it will end up on the slush pile. Also make sure that your return address is clearly written on the envelope. Follow the same procedure for all of your submissions.

Rule Nine: Make a list of all of your submissions so that you can keep track of their progress. Figure 3-2 shows a small portion of the list I used for *Black Falcon* back in 1986 and 1987. Your list should include the date of manuscript submission, the editor it was sent to, the publishing company it was sent to, the type of correspondence, the date it was returned, the date it was published (*wishful thinking*), and the amount of money spent on postage (including return postage).

Richard Cook

April 29, 1987

Erroll McDonald, Editor
Random House
201 East 50th Street
New York, NY 10022

Dear Mr. McDonald:

Having been involved with South Florida's aviation community since the 1970's, I am quite familiar with the illegal trade of aerial drug smuggling. FALCON, a 73,000 word novel, takes a fresh look at this topic from a perspective never before written about — trying to disassociate from a multi-million dollar narcotics organization once involved. The novel begins with a forceful 'power-punch' and never lets up. The subject material is authentic and timely.

FALCON's main story line follows a twenty-nine year old cargo pilot into a desperate struggle for survival against his distraught emotions, nature's elements and the deadly world of drug running. Several subplots come into play, all leading up to the novel's climactic ending.

I am enclosing a resume, sample chapters and a brief synopsis for your review. If FALCON strikes your interest, I would be most willing to send the manuscript on request.

Sincerely,

Richard Cook

ENC.

Figure 3-1

Date Sent	Editor & Related Information	Type Corres.	Date Returned	Date Published	Expenses	Payments
11/6/86	Nan A. Talese - Editor-in-Chief - Houghton Mifflin	Query Letter	1/7/87(-)	—	$ .44	—
11/6/86	Ashbel Green - Senior Editor - Alfred A. Knopf	Query Letter	11/21/86(-)	—	$ .44	—
11/6/86	M.E. Treadwell [illegible] - Managing Editor - Little, Brown Co.	Query Letter	11/21/86(-)	—	$ .44	—
11/6/86	George Walsh - Editor-in-Chief - Macmillan Publishing	Query Letter	12/12/86(-)	—	$ .44	—
11/6/86	Thomas H. Quinn - Editor - McGraw-Hill Book Co.	Query Letter	11/21/86(-)	—	$ .44	—
11/6/86	Sherry W. Arden - Publisher - William Morrow & Co.	Query Letter	—	—	$ .44	—
11/6/86	Arnold Dolin - Editor-in-Chief - New American Library	Query Letter	11/15/86(-)	—	$ .44	—
11/6/86	Robert Bender - Senior Editor - Simon & Schuster	Query Letter	11/20/86(-)	—	$ .44	—
11/6/86	Sol Stein - President - Stein & Day Publishers	Query Letter	—	—	$ .44	—
11/6/86	Michael Denneny - Senior Editor - St. Martin's Press	Query Letter	2/19/87(-)	—	$ .44	—
11/6/86	Lyle Stuart - President & Publisher	Query Letter	11/29/86(-)	—	$ .44	—
11/6/86	Dan Frank - Editor - Viking Penguin	Query Letter	12/16/86(-)	—	$ .44	—
11/6/86	David McIntosh - Managing Editor - Walker Co.	Query Letter	11/20/86(-)	—	$ .44	—
11/6/86	Jim Frost - Senior Editor - Warner Books	Query Letter	11/20/86(+)	—	$ .44	—
11/6/86	Susan Moldow - Editor-in-Chief - Dell Publishing	Query Letter	11/15/86(-)	—	$ .44	—
11/6/86	Sterling Lawrence - Managing Editor - W.W. Norton	Query Letter	12/12/86(-)	—	$ .44	—
11/11/86	Anita Diamant - Agent	Manuscript - Cover Letter	1/16/87(-) MS returned	—	$ 10.51	—
11/15/86	June Roth - Author	Thank you Letter	—	—	$ .22	—

Figure 3-2

Rule Ten: Live your life as normally as possible. Try not to think too much about your manuscript submissions, although this will be difficult for most people.

Out of the eighty-four submission packages I sent out for *Black Falcon*, I received seventy-eight responses. Of those, I received thirty-three form letter rejections, forty-two personalized rejections, and three initial positive responses (meaning that the publishers asked me to send the entire manuscript). Six publishers never even bothered to respond.

You might be wondering what the difference is between a form letter rejection and a personalized rejection. Well, psychologically there's a lot of difference. For example, look at the difference between the form letter rejection shown in Figure 3-3 and the personalized rejection shown in Figure 3-4.

Figure 3-3 shows a portion of one of the form letter rejections I received. This letter was addressed to me as Dear Writer. It went on to say that my work was being

Dear Writer:

We regret to inform you that it is no longer [illegible] policy to accept unsolicited manuscripts.

The volume was so great that we could not give each manuscript the careful attention that each deserved. We are returning your work without review at this time, and wish you the best of luck placing it with another publisher.

Figure 3-3

Dear Mr. Cook,

Thank you very much for sending me the material on Falcon, but I'm afraid your novel is not right for me. In any event, I wish you all success in placing it with another publisher.

Figure 3-4

returned without review because they no longer accepted unsolicited manuscripts. In other words, my manuscript had not even been given a chance.

Figure 3-4 shows a portion of one of the personalized rejections I received. This letter was addressed to me by name. It also mentioned my manuscript by name, letting me know that my submission package was at least looked at by someone at that particular publishing house.

For me, the high point of the submission process came when I received each of the three initial positive responses. The companies that sent them to me were Crown Publishers, Warner Books and one other New York publishing house (which shall remain nameless because they turned out *not* to be a major publisher at all, but rather a major subsidy publisher). Figure 3-5 shows the initial positive response I received from Warner Books. The one from Crown Publishers was similar. The third one was a two-page glorification of my manuscript, followed by a request for lots of money if I wanted them to publish my book (which tipped

WARNER BOOKS, INC.

666 Fifth Avenue
New York, New York 10103
212 484 2900

November 17, 1986

Dear Richard Cook:

I will evaluate FALCON if you care to send it to me.

Yours sincerely,

James D. Frost
Executive Editor
Hardcover

JF:es

Figure 3-5

me off to stay clear of this firm).

I'm almost too embarrassed to admit that when I received the initial positive responses from Warner and Crown, I thought the sale of my manuscript was going to be a done deal. I soon learned, however, that just because an author gets asked to send in the entire manuscript, it doesn't mean a deal is even close to being made. Both companies eventually rejected *Black Falcon*. The final outcome of my submission process was eighty-four rejections for *Black Falcon*.

Independent Publishing. With more than 40,000 independent publishing companies scattered across the nation, unagented and unpublished authors have a much greater chance of getting their manuscripts picked up by one of these smaller companies than by one of the major publishing houses.

For example, *Writer's Market* quoted Naval Institute Press (the independent publishing company that launched Tom Clancy's writing career) as saying that most of their titles (70% to be exact) came from previously unpublished writers.[12]

With this being true, authors can get away with submitting their manuscripts in an unprofessional manner, right? Wrong! The rules of submission are just as strict and rigid for small presses as they are for major publishers. The only difference is that unsolicited manuscripts have a much greater chance of getting read at the smaller companies. Why is this so? Because the one hundred or so superconglomerates count on bestselling authors to bring in most of their profits, while independent publishers count on everybody else. The amazing thing about all of this is that small presses are doing fantastically well operating in this fashion.[13]

What allows independent publishers to do well with manuscripts that have been repeatedly rejected by major publishing houses? Low overhead is one of the answers. In an editorial written for *Small Press* (a magazine geared towards independent publishers), John P. Dessauer explained that major publishers operate their companies with such a high rate of inefficiency, they can only earn profits from books that have high commercial appeal. Independent publishers, on the other hand, run their businesses with such low rates of inefficiency, they can earn profits from books that have narrow audiences - books that major publishers would never even consider putting into print.[14]

Whereas large publishers are not even guaranteed to reach their break-even point with 20,000 copies sold, small presses regularly earn profits on fewer than 5,000 copies.[15] Quite a number of independent publishers are even earning profits on press runs ranging from 1,000 to 2,000 copies.[16] Because small presses can earn these profits, they are able to take chances on books written by unpublished authors - books that might not sell very many copies.[17]

Another reason why independents are so successful with unpublished authors is that most of these companies are owned and operated by individuals - not corporate businessmen and stockholders. These individuals consider books to be labors of love and creativity - not some type of commodity. They treat authors as artists - not factory workers. They savor everything about the literary world! Because of this, they are only too happy to take on fledging authors, molding their writing careers into something special.

A third reason for the success of small presses is that each of these companies aims its books at specific markets. In other words, one small press might publish books about airplanes, another might publish books about boats, another might do books about gardening, and so on. You can pick any subject and I'll bet that there are at least one or two small presses that specialize in books on that subject. This *specialization* is what allows independent publishers to know which books will sell well year after year so that most of their books will never go out of print. Compare this philosophy with that of the major publishers, who take most of their books out of print after only a month on the bookstore shelves (unless a book becomes a bestseller, which is extremely rare). A good example of specialization in the small press industry is Rainbow Books, which is based out of Highland City, Florida. This independent publishing company began operations by concentrating solely on books about UFOs, child abuse, and kidnappings of children by their parents.[18]

A fourth reason for the growing success of small presses is that they promote *all* of their books and *all* of their authors - not just their bestsellers and mega-authors. Books written by first-timers get just as much publicity as the guys and gals working on their eighth book. This is not the case at the major publishing houses, where millions of dollars might be spent on the promotion of their mega-authors, while little is spent on everybody else. Promotion is very important for any author seeking to gain *name recognition* within the publishing industry.

Unfortunately, everything good must also have some bad. This is the way of the world. The small press industry is no exception. In this case, the downside is that smaller advances are paid to authors and sometimes lower royalty rates as well. Royalties, however, will usually continue for years and years into the future, making it possible to earn good money in the long run.

Another disadvantage of small presses is that even though these companies provide a better opportunity for unagented and unpublished writers, they still don't provide guaranteed publication. Thousands of authors are rejected each year by independent publishing companies just as they are from major publishing houses. Life's not fair, is it?

Vanity Publishing. Vanity publishing (also known as subsidy publishing) is not really publishing at all. In fact, many vanity publishers operate under fraudulent pretenses, claiming that they *are indeed* real publishers.

Vanity publishers advertise in magazines and newspapers all across the country, claiming that they are actively seeking authors. You've seen their ads and you've probably thought about writing to them (admit it). Don't feel bad, because most of us have thought about it at some point in time. But don't be fooled by their sleek brochures (which they will most certainly send to you if you write or call them). You see, *all* vanity publishers prey on frustrated writers. That is their gimmick. They will promise you the world, charge you an arm and a leg for their services, and then they won't deliver what they promised. They can't! Why? Because the *real publishing world* knows about them - therefore, nobody will buy books that are printed by them. This seems harsh, but it's the absolute truth.

As reported by Ronald Sullivan in *The New York Times*, there was a famous court case in which the jury awarded $3.5 million in damages to 2,200 authors in a class-action lawsuit against one such vanity press. Sullivan further described how each of the authors was charged as much as $20,000 by the vanity press to have their manuscripts edited and printed.[19]

In actuality, vanity presses rarely, if ever, edit manuscripts. This is because they know in advance that the books won't sell - except for maybe a few copies to the author's family and friends. Then, if the author decides to sell the books on his or her own, he or she has to first buy the books from the company even though the books have already been paid for. If this isn't fraud, nothing is. Simply stated, using a vanity publisher will kill your writing career before it even has a chance to get off the ground. It will also deplete your savings account.

Self-Publishing. Self-publishing involves starting up your own independent publishing company (your own small press) and taking yourself on as a client. The result is guaranteed publication. Thousands of authors have successfully done it in the past and thousands more will be doing it in the future.

I have already whet your appetite with a taste of what self-publishing can do for an author (refer to Chapter Two), but I haven't gone into great depth about how self-publishing works. Basically, self-publishing involves the art of taking control of your own writing career, rather than waiting for someone else to take control of it for you. This involves learning the terminology of the publishing industry (of which about 95% can be found within the pages of *How To Write, Publish & Sell Your Own Aviation Books* and *The Self-Publishing Manual*), then directing the path of your manuscript from final draft to bound text.

To be a successful self-publisher you do not have to do any of the actual physical labor involved with producing a book (such as typesetting, artwork, printing and binding), because this work is contracted out to specialists who cater to the publishing industry at very, very reasonable prices. Just to give you an example, the cost of printing and binding the first press run of *How To Get Published - Guaranteed* (my award-winning, 328-page manual that eventually turned into *How To Write, Publish & Sell Your Own Aviation Books*) came to roughly $3.10 per book for 1,500 copies. The cost of typesetting that book (a one-time charge) added another $1.30 per book, bringing the total production costs for the first printing up to $4.40 per book. With a retail price of $18.95 for the first printing, a good profit was guaranteed as long as I followed the golden rule of independent publishing, which is to keep the overhead low.

All that is really necessary to be a successful self-publisher is a strong belief in your writing, some perseverance, a little bit of cash and enough knowledge to instruct artists, typesetters, printers and binders on how you want your book to look. That's it. You let the contractors do the real work, while you spend your time overseeing, promoting and marketing the whole venture. This is the exact same way the major publishing companies do things - the only difference being that a self-publisher does it on a much smaller scale.

The fantastic thing about self-publishing is that all of the profit belongs to the author, thereby increasing the likelihood of someday earning a decent living as a writer (or at least a pretty good part-time living, as in my case). And let me tell you, most writers *do not* earn a decent living.

Stephen Goldin and Kathleen Sky, in their reference manual *The Business Of Being A Writer*, clarified an author's chances of becoming financially independent by saying only 10% of all authors in the United States will actually get published. Out of those who *do* get published, only a small percentage will ever earn enough money to write on a full-time basis.[20]

Quoting income figures that came from a questionnaire sent out by the Authors Guild Foundation, Goldin and Sky wrote that 50% of the authors who responded to the AGF survey brought in annual writing incomes of $5,000 or less. Of that 50%, quite a few had brought in $1,000 or less annually.[21]

By self-publishing a book, an author can keep that book in print year after year, slowly building up the number of books sold, slowly building up the dollars earned - all the while building up a publishing track record, gaining credibility and making headway toward the world of big-time publishing.

It should be apparent to you by now that self-publishing can be a lucrative venture for many authors. The real beauty, however, is not the money, but rather it's the track record that can be built up at an author's own pace. Take me, for example. Before I started Tri-Pacer Press, I lacked a track record of any kind. I was totally unpublished - not even a single piece in a high school newspaper. That all changed in 1988 when *Black Falcon* went through the publishing process. During my first five years as a published author, I built up the following track record: two books published (one fiction, one nonfiction), two major magazine articles published (one aviation-related, one non-aviation-related), three magazine editorials published (all aviation-related), one published book in reprint (fiction), one published book a national award-winner (nonfiction), one major magazine article reprinted in two other publications (non-aviation-related), plus book excerpts in several magazines and newspapers (fiction). On top of all this, I made several major guest appearances and a few dozen minor appearances at book shows, air shows, boat shows, and other special events, plus I got to know hundreds of publishing-related people on a first name basis. And that was just during my first five years as a self-published author! Things have been climbing steadily ever since!

By self-publishing, I have never been out of print as an aviation author and *never will be out of print!* This is the beauty of self-publishing and this is the purpose behind *How To Write, Publish & Sell Your Own Aviation Books*.

Of course, the money is nice, too. But money for its own sake isn't worth a dime. It's the financial freedom that money can provide that is important. For me, it is the freedom to eventually quit my job as an aircraft refueler so that I can spend more time writing. My writing income has already allowed me to quit my second job (the part-time job I held from 1989 to 1996 as a substitute teacher, which paid for my sport flying). Someday my writing income will hopefully provide me with the financial freedom to buy my own airplane, plus the freedom to sail down to the Caribbean for extended cruising.

Another advantage of self-publishing is the control an author has over his or her own books. When an author signs a book contract with a publishing house (large or small), the author's manuscript becomes the property of that publishing house. From that point on, the author loses all control over most of the decisions regarding that book, including editorial changes, title changes, cover design, layout, promotion and marketing. In contrast, when an author self-publishes a book, he or she retains all control over his or her creative endeavor. And don't think for a moment that a self-publisher can't produce a book that matches the quality achieved by the major publishers. Figure 3-6 shows a photograph of *Black Falcon* on display at a Fort Lauderdale bookstore. In the photograph, my novel is perched center-stage among the *New And Noteworthy* books at this particular bookstore. As you can see, my book was displayed right alongside the books published by the major publishing houses. No difference can be seen in the quality of my cover design compared with the other books. The same can be said for the text of *Black Falcon*. Figure 3-7 shows the first page from Chapter Eleven of my aviation novel. It is right on par with industry standards. (By the way, if you would like to purchase a copy of *Black Falcon*, please refer to the *Writer's Book Mart Catalog* in the appendix section of *How To Write, Publish & Sell Your Own Aviation Books*.)

As far as promotion and marketing are concerned, a self-publisher knows the potential market for his or her own book better than anyone else. After all, he or she is the author of the book. Therefore, a self-publisher is the most qualified person to direct the promotion and marketing efforts. Using *Black Falcon* as an example, I wrote the book as a Florida-oriented novel revolving around airplanes and sailboats. Because of the book's subject matter, I aimed my publicity campaign (review copies and press releases) at local and national aviation and sailing magazines, as well as Florida-based newspapers. My efforts resulted in several dozen book reviews, feature articles and news briefs about

New And
Noteworthy
RICHARD
BURTON
BLACK
FALCON
MUTATION
MIDNIGHT
STORMING
INTREPID

Figure 3-6

BLACK FALCON 121

CHAPTER ELEVEN

Saturday morning

Sunrise was still two hours away when Rob finally exited the airfreighter. He joined Nick, who had gone down the ladder half an hour ago and was sitting on the damp ground with his back resting up against one of the main tires. Rob plopped himself onto the ground also and leaned against the opposite side of the tire. The only sounds in the calm morning air were those of a couple of guards laughing in the distance.

"One of the better strips Gustavo's picked out, wouldn't you say?" Rob commented.

"Yeah, man, 'cept for them trees down there at the other end. We're gonna be departing right for them. Better pray God's on our side for this one."

"We've got five thousand feet, Nick."

"Yeah, man." Nick closed his eyes. He was too exhausted for any more small talk—too old to be flying thirty-odd hour missions down to South America and back. To ease the effects of his tiredness, Nick drew out a cigarette and jammed the Marlboro between his lips. As he smoked the cigarette, he let the ashes fall onto his lap, not bothering to flick them elsewhere. The chain-smoking copilot had already finished off his first pack of

Figure 3-7

Black Falcon and Tri-Pacer Press.

Where has all this gotten me? Well, let's just say that *How To Write, Publish & Sell Your Own Aviation Books* tells the true-life story of my first five years as a self-publisher (my novice years, so to speak). During those first five years, self-publishing put me onto a course fringing along the border of the major publishing industry. *(I know in my heart that someday I will cross that border!)*

Am I angry that the major publishing industry never took a chance on *Black Falcon*? I used to be, but not any more. Now I fully understand why the major publishers can't afford to take chances on unproven authors. Fortunately, I had the support of the self-publishing industry to help me prove my worth as a writer. That's right - I had the support of the self-publishing industry behind me 100%. And I'm not alone. For example, Dan Poynter, the author of *The Self-Publishing Manual*, got his start as a self-published aviation author by writing a guide to sport parachuting. He had some copies printed up, then sold them at skydiving centers throughout California (hence, the name of his publishing company, Para Publishing). Poynter's first book was such a huge success that he took it back for several reprintings, and it is still in print today. In 1973, he wrote and self-published his second book - this one about hang gliding. Would you be impressed if I told you that Poynter has since sold more than 125,000 copies of this bestseller?[22]

Because of Poynter's success, he decided to share his knowledge with other authors by writing what has since become his masterpiece, *The Self-Publishing Manual*. This book is now considered to be the ultimate reference guide for self-publishers, whether they are working on their first book or their tenth. Why do I keep mentioning this book? Because it has been the primary foundation behind the success of so many self-publishers, including myself. (To order a copy of *The Self-Publishing Manual*, please refer to the

Writer's Book Mart Catalog in the appendix section of *How To Write, Publish & Sell Your Own Aviation Books.*)

By the way, self-publishing is nothing new, although it has become more common with the advent of personal computers. Probably one of the most famous self-publishers of all time was Mark Twain. This name is familiar to almost everybody, but what most people don't know is that Mark Twain self-published his American classic *The Adventures Of Huckleberry Finn*. That's right - Mark Twain was a self-publisher during a portion of his writing career, as were many other world-renowned authors.

Twain's first book, *The Celebrated Jumping Frog Of Calaveras County And Other Sketches*, was rejected by all of the major publishing houses of the day. A friend offered to print the book for him - sort of a self-publishing type arrangement, but with the self-publisher being his friend. Each man agreed that Twain would receive 10% royalties, but he never received a cent. Basically, Mark Twain was ripped off by his friend.[23]

Shortly thereafter, Twain's second book, *The Innocents Abroad*, was picked up by the American Publishing Company for 5% royalties. Twain's third book was also published by this company. The year was 1872. Over the next decade, Twain signed over several more books, all with 71/2% to 10% royalties. At the end of the decade, Twain discovered that he had been swindled out of a good portion of his royalties from each of his books.[24]

Disgusted with the major publishers, Mark Twain paid a subsidy publisher (another name for a vanity press) to produce his next two books, *Old Times On The Mississippi* and *The Prince And The Pauper*. Both of these attempts at publishing were failures. That's when he decided to self-publish his next book, *The Adventures Of Huckleberry Finn*. The year was 1884.[25] The book went on to become a literary classic (a true masterpiece) and it is still in print today - more than one hundred years after Twain originally self-published it.

Is self-publishing worth it? Well, just ask Denton Rickey Moore, a self-publisher who lives just north of Seattle, Washington. Although he's not an aviation author, he has an excellent track record as a self-publisher. I met Denton at the BookFest of the Palm Beaches in April 1993. I had a booth at one end of the book show, while he had a booth at the other end. Both of us were busy selling autographed copies of our respective books, but neither one knew the other was at the show. Well, when a customer of mine showed me a copy of Denton's book, I knew I had to go introduce myself to him. You see, Denton Rickey Moore was a sailor. His nonfiction book, titled *Gentlemen Never Sail To Weather: The Story Of An Accidental Odyssey*, told about his five-year circumnavigation aboard a 50-year old wooden sailboat (which happened to be one of my long-term goals at the time).

I walked over to Denton's booth, introduced myself, told him how I found out about his book, then told him about my air-sea adventure novel. We talked for a couple of minutes about our respective books, then exchanged autographed copies.

Denton then showed me a double-sided brochure he had printed up for his book. Figures 3-8 and 3-9 show both sides of this brochure. (A brochure like this would be great for a true-life, aviation travel book.) On the top of the brochure was the statement Book Of The Month Club Selection. I asked him about it and he said that a representative from Book Of The Month Club had seen *Gentlemen Never Sail To Weather* at the COSMEP booth at the previous year's American Booksellers Association trade show (now called BookExpo America) and had contacted him to make a deal. I was impressed!

I was further impressed when Denton told me that he

Figure 3-8

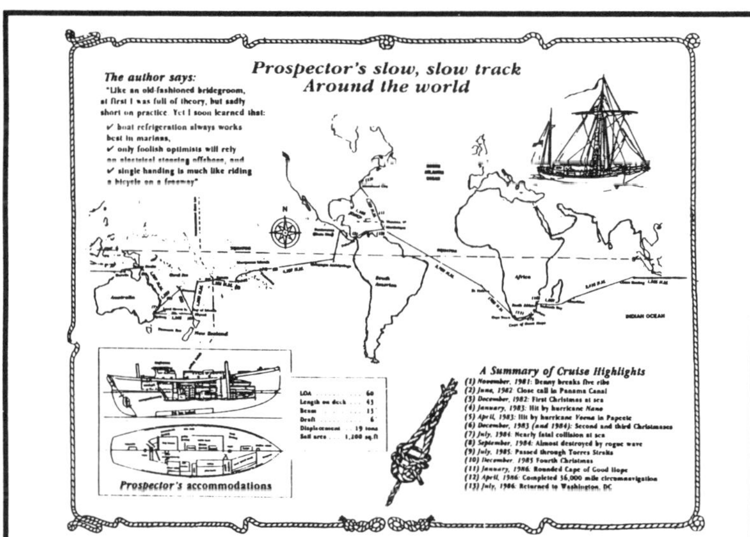

Figure 3-9

had sold out his first printing of 2,000 hardcover copies in less than a year, then released a second printing in a trade paperback edition. I really liked his style. He has since self-published his second book, *Alaska's Lost Frontier*, and is now publishing books written by other authors.

So now you have a feel for what can be accomplished when ordinary people self-publish their books. And it's not just a few people. Thousands of authors have successfully self-published their own books. According to statistics that were printed in *The New York Times*, just over 1,000 authors self-published their books in 1981 (the year *Black Falcon* was originally developed into a book idea). The statistics for self-publishers doubled to just under 2,000 authors by 1987 (the year I decided to become a self-publisher), and this figure jumped to well over 5,500 authors annually by 1995 (the year after *How To Get Published - Guaranteed* won a book award from the National Association of Independent Publishers).[26] I'd like to think that my frequent lecturing during that period (as well as the availability of my how-to book) had something to do with the increasing number of authors becoming involved with self-publishing. Maybe it did, maybe it didn't. I have no way of finding out, but in either case, the increase happened for real and it's still happening today.

How much success can the average self-publisher hope to achieve? Well, many of us hope to eventually sell our books to major publishers, but that doesn't happen often - although it does happen occasionally. For example, a novella titled *The Christmas Box*, which was self-published by Richard Evans in 1992, was sold to Simon & Schuster in 1995 for $4.2 million.[27] A more realistic goal is to sell out the first printing of a book and go back to press, thereby building a solid publishing track record for future books.

For those of you who still aren't quite sure about self-publishing's place within the publishing hierarchy, let me give you two analogies from within the aviation industry - the first being air transportation companies and the second being manufacturers of general aviation aircraft.

The top of the air transportation hierarchy includes the national and international air carriers, which fly to major destinations (using equipment such as A-300s and MD-80s). These airlines are analogous to major publishers. The next lower position in this hierarchy includes the regional air carriers, which fly to smaller cities within specific areas of the country (using smaller equipment such as DHC-7s and EMB-110s). These companies are analogous to small publishers. The bottom rung of this air transportation hierarchy (but of no less importance) includes the many private companies that fly on-demand charters to wherever their customers want to go (using even smaller equipment such as Learjets and Challengers). These firms are analogous to self-publishers. Are you now getting a feel for the true legitimacy of self-publishing?

This same type of hierarchy can be found within the general aviation aircraft manufacturing industry. The top of this hierarchy includes the manufacturers of most of the general aviation aircraft that are found on today's airport ramps (companies such as Cessna and Piper). These manufacturers are analogous to major publishers. The next lower position in this hierarchy includes the manufacturers of specialty-type aircraft that are not seen in great numbers on today's airport ramps (companies such as Bellanca and Maule). These manufacturers are analogous to small publishers. The bottom rung of this aircraft manufacturing hierarchy (but of no less importance) includes the thousands of homebuilders who construct aircraft in their garages and hangars from kits or plans (such as the popular Kitfox and Lancair homebuilts). These aircraft homebuilders are analogous to self-publishers. As a matter of fact, the aircraft homebuilding movement has been growing just as rapidly as the self-publishing movement - so much so that 18% of today's single-engine piston fleet is comprised of homebuilt aircraft.[28] This proves without a doubt that grassroots movements within the United States have become readily acceptable and legitimate in today's complex society.

If you believe that an author has every right to promote and market his or her own writing, and if you enjoy testing the limits of your creativity and ingenuity, then self-publishing just might be the break you've been looking for - the break that will lead you toward guaranteed publication. It's your choice!

CHAPTER FOUR

REASONS FOR WRITING

Why do people become authors? Why do they punish themselves day after day with the lonely process called writing, facing a blank sheet of paper or perhaps a computer screen, knowing that they must fill those spaces with words of wisdom, fact, humor, knowledge, suspense or adventure? Why do they face the grind of it all? Aren't there already enough books written on every subject imaginable? Why do people become authors?

The only true answer is that every author has his or her own reasons for wanting to write. It is very possible (and common) for those reasons to be multi-faceted or even to change over time. To give you a clearer understanding of how an author's mind works, follow along as I take you back in time to 1981, and then forward again, through the meandering journey that has been my writing career. It's definitely not a boring ride and you might just see a bit of yourself as you read about my evolution into authorship.

Back in 1981, I had some problems that seemed to hover over me like a ton of mud. I couldn't get a grasp on life, going from one day to the next without any particular direction in mind. I was twenty-three years old and living a typical blue-collar existence fueling corporate aircraft at Fort Lauderdale-Hollywood International Airport. All I really enjoyed doing was hanging out around airplanes. I liked living in South Florida, but the area's population (which reached well into the millions) held no importance for me.

In short, I was a loner searching for something, but not knowing what it was I was searching for.

Then I found it! Novels! Aviation adventure novels in particular! That was extremely ironic because in all my years of schooling, I had never read fiction for pleasure. You see, the novels that had been forced upon me as required reading during high school had never appealed to me. Therefore, I came to the conclusion that all novels were dull. Sure, I had read lots of aviation technical books and magazines, but I didn't even know that aviation novels existed.

So how did aviation novels suddenly become the answer to my search? It all started one day during the early part of 1981 when I stopped in at a Fort Lauderdale pilot shop to browse. It was there that I spotted an aviation novel titled *The Pilot* by Robert P. Davis. The cover said it was a tale about an alcoholic airline pilot. I bought the book partly out of curiosity, partly out of boredom, not knowing whether or not I was going to enjoy reading it. What I found out was that I couldn't put the book down. I devoured its 276 pages within days.

The characters, the flying sequences, the emotions in that fast-paced, action-adventure novel all seemed so real to me. Over the next couple of months, I raced through dozens of other aviation-oriented novels, including many by the late Ernest K. Gann (who passed away on December 19, 1991), including *Island In The Sky, The High And The Mighty* and *Band Of Brothers*, as well as Gann's non-fiction classics, such as *Fate Is The Hunter* and *Ernest K. Gann's Flying Circus*. I was hooked on reading!

During the summer of 1981, I purchased a copy of *A Hostage To Fortune*, Ernest K. Gann's autobiography. That book changed my life. Within its pages I found a purpose in life, a path to follow, a goal to strive for, a new beginning. I was inspired by Gann's career as a pilot, sailor and author. He lived adventure! I decided right then and there that I was going to emulate his lifestyle - but who was I kidding? I had no formal education in writing, nor did I know how to plot a book. In fact, grammar and composition were two of my worst subjects in school. Still, despite those big handicaps, I played around with the idea of writing an aviation-oriented novel. Sometimes I would convince myself that I could indeed write the book, while at other times I would disappoint myself with discouraging thoughts of failure.

Then fate came into play late one night towards the end of 1981 when the entire plot for a novel popped into my head, complete from start to finish. I got up out of bed and wrote down the basic story line for *Falcon* (later to become *Black Falcon*), an action-adventure novel revolving around South Florida's deadly narcotics trade. From that point on, there was no turning back. *Black Falcon* had become my child and it was up to me to nurture it and feed it and bring it to its fullest potential. I was going to be a novelist and that was that!

Originally, I had two distinct reasons for wanting to write *Black Falcon*. First, I saw it as a form of therapy - a means to finally solve my problems by putting those problems into the lives of the various characters in the story. I figured that this would allow me to see my life from the perspective of an outsider looking in. Secondly, I really *did* want to emulate Ernest K. Gann's lifestyle of writing, flying and sailing (even though I knew nothing about sailing at that time, except what I had read about it in Gann's autobiography). I set my goals high right from the start.

The first step in attempting to write *Black Falcon* was an expansion of the plot from a single page of ideas into an eight-page outline. This brief outline was then stretched into a thirty-page, chapter-by-chapter outline. Next came decisions about locales, the development of three-dimensional characters and the figuring out of motives and methods within the story line.

I wanted my novel to be realistic, so there was one more step I had to undertake before actually writing the book. That step was research. Gann had written in his autobiography about the importance of an author knowing his subject expertly before trying to write about it. He explained that this was because readers were not easily fooled by phony authenticity.[29] This expertise can only be obtained through intense research. Therefore, I threw myself into four months of solid research. I found this stage of the writing game to be exciting and challenging, with the research taking me on jaunts to the Florida Keys, to a fishing camp in the Everglades, to a couple of very seedy taverns, to several South Florida airports, to a small farming community bordering along the shores of Lake Okeechobee, to several rundown boatyards, to a luxury marina, and to Gun Cay in the Bahamas aboard a 76-foot Nova Scotia schooner (christened *Freedom*).

The aviation portion of my research was fairly easy because I had an extensive aviation background as a sport pilot, aircraft refueler, amateur aviation history buff and all-around airport bum. I had access to lots of pertinent information, such as aeronautical charts for flight planning data, aircraft specification sheets, flight manuals and aircraft performance figures.

Because I lived in South Florida, researching the drug smuggling portion of my novel was also easy. I had access to information from U.S. Customs agents, U.S. Drug Enforcement personel, U.S. Coast Guard officers, police officers from various local units, plus all the wild rumors that circulated among the airport employees at many of the local airports in South Florida during the early 1980s. This information was then backed up with a newspaper subscription to *The Miami Herald*, which ran articles about drug smuggling on an almost daily basis.

The Achilles' heel of my research, however, was sailing.

You see, until I had read about Gann's sailing adventures in his autobiography, I had been oblivious to the thousands of sailboats plying Florida's waterways. I just never gave the vessels much thought. Once I became aware of sailing, however, I knew that I was missing out on a mystical way of life (*the peaceful way*) and it became part of my goal to learn about sailing and to make my book an air-sea action novel instead of just a pure aviation action novel.

Before I could do this, though, I had to first learn the terminology of sailing and gain a feel for the sea. I began by hanging out at local boatyards, striking up conversations with sailors, reading books and magazines about sailing, and collecting brochures on different types of sailboats. I also attended the Fort Lauderdale Boat Show.

The ultimate research experience occurred in November 1981 when I struck up a fast friendship with a man named George Pagar. George was the owner and captain of *Freedom*, a 76-foot wooden, gaff-rigged schooner that hailed from Boothbay, Maine. George had motored down the Intracoastal Waterway to Fort Lauderdale in order to carry out minor repairs on his sailboat before embarking on a planned round-the-world voyage. I offered to help with the repairs, thereby learning firsthand about life aboard a wooden vessel. In exchange for my voluntary help, I was invited to participate as a working crewmember on the first leg of the voyage, which was from Fort Lauderdale to Gun Cay in the Bahamas. Figure 4-1 shows a photo of *Freedom*.

After arranging a week's vacation from my job at the airport, I departed aboard *Freedom*, bound for Gun Cay across the Gulf Stream. As a crewmember, part of my responsibility was to hoist, trim and furl the cumbersome gaff-rigged mainsail. As a student of sailing who was willing to learn as much as possible in the shortest amount of time, I got a chance to take the helm as we made headway towards the Bahamas.

Figure 4-1

Several chapters could be filled about the research I did on the subject of sailing, but that would detract from this book. Let me just say that I learned enough about sailing to addict myself to the sport. Within a year of my Bahamian cruise aboard *Freedom*, I took out a $2,000 bank loan to purchase a sailboat of my own - a beat-up, run-down and faded 22-foot sloop that I christened *Dreamer*. After reading a couple of books about boat restoration, I fixed her up into a nice coastal cruiser and spent many weekends learning the finer points of sailing. Figure 4-2 shows *Dreamer* after restoration. (Note: I had to eat tuna fish and bologna sandwiches for a long time in order to pay back the loan, but it was worth it because I owned *Dreamer*.)

Did I learn enough about the sport of sailing to write a novel about it? That question was answered for me years later when *Cruising World*, one of the world's most widely-read sailing magazines, printed the following review about *Black Falcon* in their June 1989 issue: "The story is memorable and hard to put down. Sailors and fliers will love it."[30]

Figure 4-2

As I added page after page to my research notes, I knew deep down that my novel was going to be authentic. Figure 4-3 shows some of the research material I used, including nautical and aeronautical charts, newspaper and magazine articles, books, brochures, and various other items related to airplanes, sailboats and drug smuggling.

I was itching to get started with the actual writing of the book, but I held off until February 1982 when my research was completed. At that time, I sat down with a pen and paper (no word processor or computer for me back then) and began writing out the first draft, making sure to work my personal problems into the lives of the characters.

After nearly a year-and-a-half of torturous sessions (and that's exactly what they were), the first draft of *Black Falcon* was not even half finished - but I had solved my problems. Unfortunately, that meant that I had lost one of my main reasons for wanting to write the novel, which was to solve my problems. With my problems gone, my outlook on life improved immensely and I started spending more

Figure 4-3

time working on my sailboat, hanging out at boatyards, sailing, hanging out at airports (I wasn't flying at the time due to a money shortage) and basically pushing my writing off to the side. One day I just packed up the manuscript and stored it in my closet, where it stayed for several years. My dream of becoming a novelist was put on hold for awhile so that I could live the reality of adventure, rather than just writing about it. During that period, I also got married.

One year into my marriage, I unearthed a new reason for wanting to complete my novel. On a visit to the South Street Seaport in New York City during the summer of 1985, I learned that George Pagar, the captain of *Freedom*, had died of cancer. Because he had made such a powerful impact on my life by showing me the freedom and beauty of the sea, I decided to finish writing *Black Falcon* as a memorial to him. Upon returning to Florida, I dusted off the manuscript and proceeded with the first draft. The more I read it, however, the more I disliked it because it was too full of my old personal problems, so I threw it away and started over from scratch - although the basic story line remained the same.

Once I got seriously into the writing, one of the reasons I had originally had for wanting to become an author came back to me - that of emulating the lifestyle of Ernest K. Gann as a pilot, sailor and author. It was this reason that later led me to create the outlines for two sequels to *Black Falcon* (titled *Death Wreath* and *Avenge*), which will make this a trilogy revolving around my main character, a pilot named Rob Jensen.

Almost a year after beginning the rewrite of *Black Falcon*, the manuscript was finally completed. The year was 1986. My goal of creating a memorial to George Pagar had been accomplished. The novel was then rejected by eighty-four publishers between 1986 and 1987. I formed Tri-Pacer Press in 1988 and self-published *Black Falcon* that same year. By 1989 I was financially able to start sport flying again. Thus, my goal of living the adventurous lifestyle of Ernest K. Gann as a pilot, sailor and author was successfully reached. Today, my personal business card reads: Author, Pilot, Sailor, Adventurer, Thrill-Seeker.

In short, *Black Falcon* was written because of three very diverse reasons - the need to solve personal problems, the desire to live a certain lifestyle, and the need to pay homage to a man of courage. All three turned out to be excellent reasons for wanting to write. Without them, I wouldn't be an author today.

I can't tell you what your reasons should be for wanting to write a book. Whatever they are, though, keep them in the forefront of your mind and let them inspire you into authorship.

CHAPTER FIVE

STEP ONE: WRITE YOUR BOOK

You cannot guarantee the publication of your book without first writing your book - and believe me, writing your book is the hardest part of self-publishing.

Even though I have written and published fiction and nonfiction, I do not profess to have any magical formula for writing books. I have never taken any creative writing classes, nor have I ever attended any writing seminars (except as a guest speaker, which is ironic). I am terrible when it comes to grammar - especially when it comes to commas, colons, semi-colons and all the rest of that stuff. I wouldn't be able to describe the difference between a verb and an adverb if my life depended on it. So with all of this riding against me, how have I managed to find success in the writing trade? Easy! I write like I talk. In other words, I write what sounds natural to me.

While writing my books and magazine articles, I have developed some very definite opinions about the craft of writing that I would like to share with you. I have laid these opinions out for you in great detail over the next several pages. Hopefully, these opinions will make it easier for you to write your own book.

Writing is hard work, so don't ever let anyone tell you that it's not. And because it's so difficult, every author should have a source of encouragement. I found my initial encouragement in a series of letters written to me by a cousin of mine, George Diamond. Throughout the years that I spent writing *Black Falcon*, George's letters always seemed to give me the spark that I needed to keep going.

A second source of encouragement came from a book I received from my brother, Steve Cook, back in 1982, titled *Maybe You Should Write A Book* by Ralph Daigh. I don't know whether or not Daigh's book is still in print after all these years, but if not, an equally fine book that can be used by you as a source of encouragement is the bestseller *Is There A Book Inside You?* by Mindy Bingham and Dan Poynter. (To order a copy of *Is There A Book Inside You?*, please refer to the *Writer's Book Mart Catalog* in the appendix section of *How To Write, Publish & Sell Your Own Aviation Books*.)

Once an author finds his or her source of encouragement, the next step is to figure out what type of book to write. Will it be fiction or nonfiction? Will it be commercial or literary? A book is usually started from one or two initial ideas. As I mentioned in the previous chapter, the initial idea for *Black Falcon* came to me one night back in October 1981. The published version of this book is basically just an expanded version of the initial idea.

An initial idea for a book can be either a new idea or a new approach to an old idea. For example, *Black Falcon* was the latter - a new approach to an old idea. Hundreds of novels had been written about the drug smuggling trade, but all of them had taken the point-of-view of the cops versus the smugglers. My novel, however, took the point-of-view of the smuggler pilot versus the smuggling organization. My initial idea was this: What if an airline pilot was offered $100,000 cash to make one drug flight from Colombia to Florida, then changed his mind and tried to disassociate himself from the smuggling organization? It's a simple idea and it worked. In fact, simple ideas are usually the ones that *do* work. Whether a book is an aviation novel, an aviation textbook or a how-to manual about building an airplane, the initial idea should be a simple one. If the idea cannot be described in one sentence, it is probably too complicated.

A good book idea should also revolve around subject matter that the author knows about intimately - or at least something that he or she plans on learning about before writing the book. For example, with *Black Falcon* I already had firsthand knowledge about airplanes and Florida, and I knew I could get access to information about smuggling, but I had to go the extra mile to learn about the sport of sailing. Another good example is *How To Write, Publish & Sell Your Own Aviation Books*. I did not have the knowledge to write this book in 1988, but by 1994 I had gained enough firsthand experience in the field of self-publishing to put out *How To Get Published - Guaranteed*, which was then transformed into *How To Write, Publish & Sell Your Own Aviation Books* in 1998. One thing I've come to realize is that an author has a much greater chance of creating a successful book if the subject matter is about something he or she is familiar with and enjoys.

A third characteristic of a book idea is that it can come from anywhere or anyone. For example, the initial idea for my second book, *How To Get Published - Guaranteed*, was not my own idea. I had been giving a talk to a group of Palm Beach authors back in 1990 when one of them suggested that I should write a book about my experiences with self-publishing. Up until that time the idea had never even entered my mind, but when everyone in the audience agreed that a book about how I successfully self-published *Black Falcon* would be a big hit with unpublished authors, I realized they were probably right. It wasn't until April 28, 1991, however, that I actually acted upon the idea and began jotting down notes for the various chapter headings.

Once I started *How To Get Published - Guaranteed*, the ideas kept flowing - just like they did during the early developmental stage of *Black Falcon* ten years earlier in

1981. I call this stage of the writing process the idea notes stage. During this period, ideas seem to pour into my head on a 24-hour-per-day basis. Any idea that is not written down will likely be lost. Because of this, I try to keep a pen and a pad of paper with me at all times while in the idea notes stage of writing (whether it be for a book or a magazine article). Some authors might prefer using a small tape recorder for this purpose, which is perfectly fine, especially when you consider that on the few occasions when I've been caught without a pad of paper, I've had to resort to using scraps of paper, pieces of cardboard, napkins, paper towels or whatever else was handy at the time. Some of these makeshift notes can be seen in Figure 5-1, which shows a sampling of actual idea notes from *How To Get Published - Guaranteed*.

Again, I would like to stress the importance of writing down idea notes the moment they burst into your mind. You could be in the shower, or driving a car, or even sleeping when an idea hits you. Whenever this happens, try to force yourself to stop whatever it is you are doing in order to write down your idea notes. If you don't, there's a good chance you will lose the thought.

As you collect your idea notes, save them in a file folder and store them in a safe place. The time will soon come when you will need them to begin writing your manuscript.

The next stage of the writing game involves organizing your book. This can be accomplished either with an outline or with a table of contents. I used an outline for *Black Falcon* and a table of contents for *How To Get Published - Guaranteed* and *How To Write, Publish & Sell Your Own Aviation Books*. The purpose of this phase is to break down the book into separate chapters, with each chapter becoming its own entity (with a beginning, middle and end). It is far less scary to think about writing one chapter at a time than it is to think about writing an entire book all at once. In

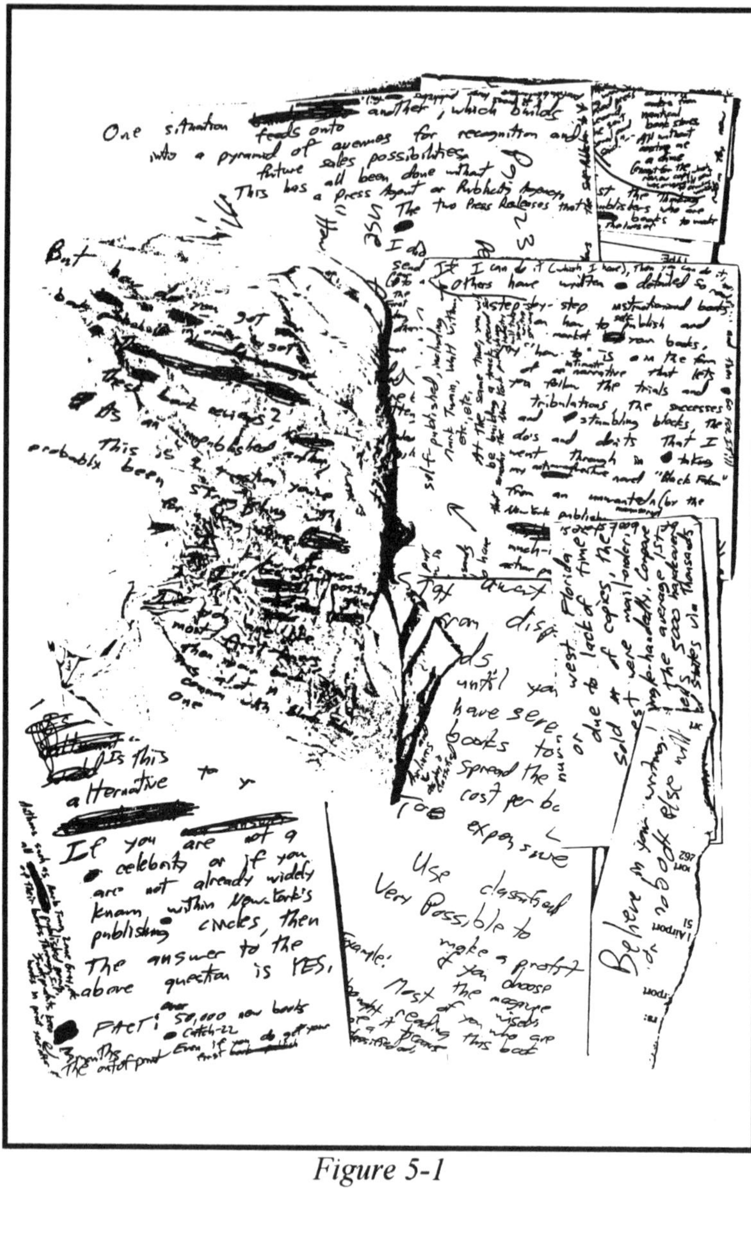
If you are not a celebrity or if you are not already widely known within New York's publishing circles, then The answer to the above question is YES.
Use classified
Very Possible to make a profit
Believe in your writing!

Figure 5-1

other words, try to imagine that each chapter is a five to twenty-five page short story, rather then frustrating yourself with the task of writing a 250-page book. This method of organization really works!

An outline doesn't have to be lengthy in order to be effective. Figure 5-2 shows one page from the outline of *Black Falcon*. On this particular page, I described what happened in Chapters Eight, Nine, Ten and Eleven. When I organized *How To Write, Publish & Sell Your Own Aviation Books*, I simply used a table of contents like the one at the beginning of this book.

The next stage of writing a book involves organizing your idea notes. To do this, pull out your file folder and rewrite all of your idea notes in a spiral notebook, being sure to skip a line between each different note. Figures 5-3 and 5-4 show two pages from the spiral notebook I used for *Black Falcon*. Figure 5-3 depicts the cockpit layout of a Douglas DC-4 instrument panel (the main airplane in the novel). I drew this sketch while sitting inside a DC-4 at Fort Lauderdale-Hollywood International Airport, all the while gathering a feel for the derelict airfreighter, its musty odor, its worn leather seats, its grime and grease from years of hard use. Figure 5-4 lists some of the idea notes I wrote down while doing research on fishing in the Everglades and Lake Okeechobee (important details since the main character in the novel is an avid fisherman). All together, my spiral notebook had more than 100 pages of idea notes for *Black Falcon*.

After all of your idea notes are organized and you have finished doing research, it is time to start writing the first draft (rough draft) of your manuscript. Remember, write one chapter at a time. Don't bog yourself down with the entire book all at once. As you finish each chapter of the first draft, make a copy and place it in a safe place, such as a bank vault, then go on to the next chapter. Do not worry

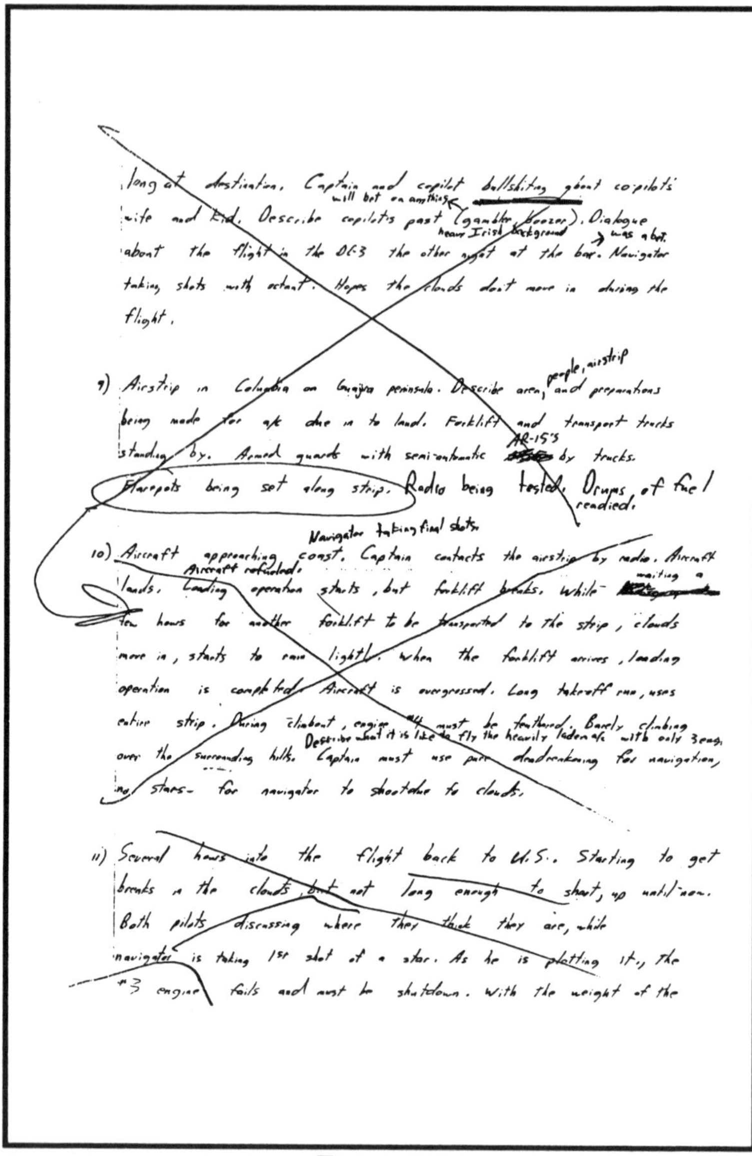

long at destination. Captain and copilot bullshiting about copilot's wife and kid. Describe copilot's past (gambler, boozer). Dialogue about the flight in the DC-3 the other night at the bar. Navigator taking shots with octant. Hopes the clouds don't move in during the flight.

will bet on anything

heavy Irish background

was a bet.

9) Airstrip in Columbia on Guajira peninsula. Describe area, people, airstrip and preparations being made for a/c due in to land. Forklift and transport trucks standing by. Armed guards with semi-automatic AR-15's by trucks. Flarepots being set along strip. Radio being tested. Drums of fuel readied.

10) Aircraft approaching coast. Navigator taking final shots. Captain contacts the airstrip by radio. Aircraft lands. Aircraft refueled. Loading operation starts, but forklift breaks. While waiting a few hours for another forklift to be transported to the strip, clouds move in, starts to rain lightly. When the forklift arrives, loading operation is completed. Aircraft is overgrossed. Long takeoff run, uses entire strip. During climbout, engine #4 must be feathered. Barely climbing over the surrounding hills. Describe what it is like to fly the heavily laden a/c with only 3 engs. Captain must use pure deadreckoning for navigation, no stars for navigator to shoot due to clouds.

11) Several hours into the flight back to U.S. Starting to get breaks in the clouds, but not long enough to shoot, up until now. Both pilots discussing where they think they are, while navigator is taking 1st shot of a star. As he is plotting it, the #3 engine fails and must be shutdown. With the weight of the

Figure 5-2

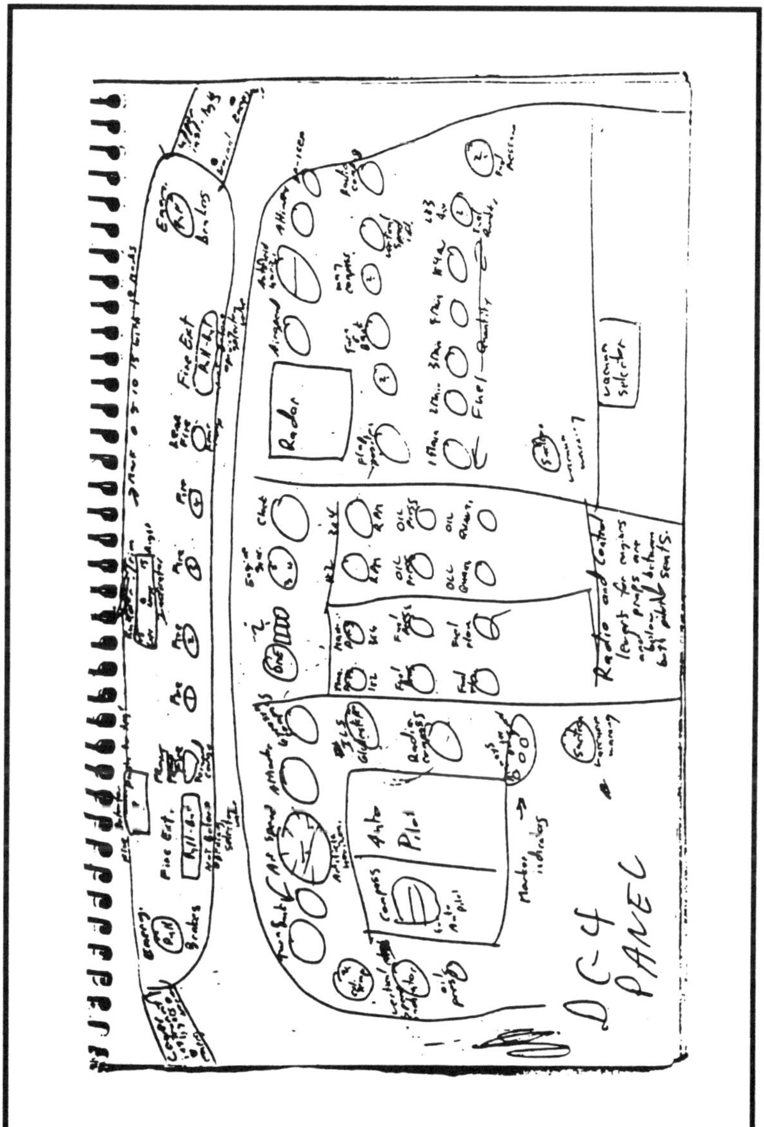

Figure 5-3

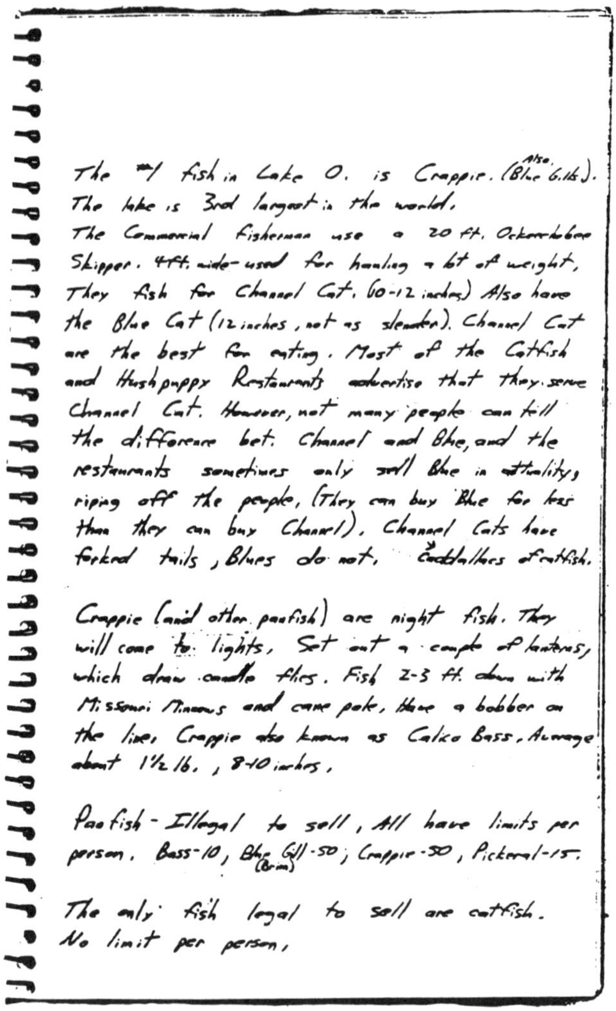

The #1 fish in Lake O. is Crappie. (Also Blue Gills).
The lake is 3rd largest in the world.
The Commercial Fishermen use a 20 ft. Okeechobee Skipper. 4 ft. wide - used for hauling a lot of weight.
They fish for Channel Cat. (10-12 inches) Also have the Blue Cat (12 inches, not as slender). Channel Cat are the best for eating. Most of the Catfish and Hushpuppy Restaurants advertise that they serve Channel Cat. However, not many people can tell the difference bet. Channel and Blue, and the restaurants sometimes only sell Blue in actuality, ripping off the people. (They can buy Blue for less than they can buy Channel). Channel Cats have forked tails, Blues do not. [illegible] of catfish.

Crappie (and other panfish) are night fish. They will come to lights. Set out a couple of lanterns, which draw candle flies. Fish 2-3 ft. down with Missouri Minnows and cane pole, have a bobber on the line. Crappie also known as Calico Bass. Average about 1½ lb., 8-10 inches.

Panfish - Illegal to sell. All have limits per person. Bass-10, Blue Gill (Brim)-50; Crappie-50, Pickerel-15.

The only fish legal to sell are catfish. No limit per person.

Figure 5-4

about editing at this stage of writing. If you've done your research well, the actual writing of the first draft shouldn't take long. Fiction, however, may take a bit longer than nonfiction (at least that's the way it's been for me). It all depends on how fluid your imagination is at the time.

When you have finished writing the final chapter of your manuscript, take some time to congratulate yourself for a job well done! After all, writing a book is a big accomplishment!

At this stage of the game, remember to always (and I mean *always*) keep a copy of your manuscript in a vault. Why do I keep stressing this? Well, I'll tell you why. Consider for a moment the possibility of disaster striking your home just after you've finished writing your book. It could be a fire, flood, earthquake, hurricane, tornado or some other type of disaster. You would be totally disheartened if the only copy of your manuscript was destroyed in a disaster - and don't be fooled by thinking that disasters only happen to the other guy or gal! I'm telling you from personal experience that disaster can strike anywhere at anytime. I learned this lesson on August 24, 1992, when Hurricane Andrew devastated South Florida in the nation's worst natural disaster, causing more than $30 billion in damages and leaving a quarter of a million people homeless. I was very fortunate that my home only received minor damage, but dozens of my friends lost everything they owned - clothing, furniture, photographs, appliances, cars, boats, airplanes, as well as their homes (*everything*). If my home had been located just twenty-five miles further south, I would have lost everything, too, including much of the research material for *How To Get Published - Guaranteed*, which I had been working on at the time. But because I had stored a copy of each of the finished chapters in my bank vault, I was assured of not losing the entire book.

Listen up, folks! It was South Florida that had a major

disaster in 1992. During the following year it was the Midwest with its flooding, followed by California with its wildfires and a killer earthquake. How many authors lost their manuscripts in those disasters? Now consider that every year there are many new disasters. Some destroy thousands of homes in one shot, while others only destroy one or two at a time. Be prepared! Don't let a disaster wipe out your writing career. Get yourself a vault and use it!

Now I'd like to pass along a few suggestions to those of you who *really do* want to write an aviation book. If it's a nonfiction book that you are planning to write, I only have one simple suggestion to help you on your way, which is to learn about your subject as expertly as you possibly can and then just write about it. This is what I did with *How To Write, Publish & Sell Your Own Aviation Books*. I wrote *Black Falcon*, self-published it and marketed it. This gave me all the expert background information I needed to write the book you are now reading. For those of you who are planning to write fiction - well, you're in for a tougher time. This observation comes from personal experience. After all, I did write *Black Falcon*, didn't I? But don't fret, because I have several suggestions to help you on your way, too. Keep in mind that people who write fiction have to create a realistic, yet imaginary world. This is very hard indeed, especially if you want your readers to believe that what you have created *is* real, even though they know it is fiction.

The most important aspect of creating realistic fiction is the development of believable characters. To do this, you must give each character his or her own identity, including thought patterns, quirks, mannerisms, body language and speech patterns. You must be consistent with these traits throughout the entire book. For example, Rob Jensen, the main character in *Black Falcon*, was raised in Florida's back-country. He was never educated past high school (except for flight instruction) and he was never exposed to the ways and means of high-society. Because of Rob's blue-collar background and lifestyle, I had him drinking beer rather than fancy cocktails whenever he was in a bar. That's just one of hundreds of examples that are found within the pages of *Black Falcon*, each one clearly defining Rob's individuality. And not just Rob. Each character in my novel has his or her own identity. No two people are alike in my book. You must do the same if you want your fiction to be good.

The next most important aspect of creating realistic fiction is called locale. This involves setting your chapters in places that are exciting, glamorous, dangerous, gloomy, exotic, etc. - places that the average person wouldn't ordinarily go to. Again using my novel as an example, *Black Falcon* is a fast-paced thriller that takes readers from a sleazy Miami bar to a salty Key West tavern, from the cockpit of a drug plane to a clandestine airfield in the mountains of Colombia, from a tropical Bahamian cruise aboard a private sailboat to a climactic speedboat chase on Fort Lauderdale's Venice-like canal system. It's all very realistic, very authentic - yet it is fiction. (Hey, does this novel sound like something you'd like to read? Remember, you can order a copy of *Black Falcon* by referring to the *Writer's Book Mart Catalog* in the appendix section of *How To Write, Publish & Sell Your Own Aviation Books*.)

Another important aspect of creating realistic fiction is realistic-sounding dialogue. Let's face it, everybody does not talk the same. You must show this fact in your novel. How? By having each character speak according to his or her background, lifestyle and individuality. For example, one of the characters in *Black Falcon* came from Colombia and never learned to speak English very well. It would have been unrealistic for me to have this character ask, "How was your flight?" Instead, I had him ask, "How flight was?" While I'm on the subject of dialogue, let me say that you should *never* add dialogue to your manuscript just to fatten up your novel. Every time one of your characters speaks, it should be for a clear-cut reason. All dialogue should be a meaningful part of the story - not filler.

Other elements of creating good fiction include suspense, conflict and a good versus evil force. A love interest can also be included in fiction, but sexually-oriented scenes should only be written into the story line if the situation calls for it. In *Black Falcon*, the situation called for it - a lot!

I do not claim to be a writing instructor, nor do I have firsthand knowledge about any of the hundreds of creative writing programs offered across the United States. I do, however, have firsthand experience writing both fiction and nonfiction books, as well as nonfiction magazine articles and editorials. From this firsthand experience, I can tell you that if you want to succeed at this writing game, you must write, write, write and rewrite, which brings us to the next chapter on editing.

(Note: How long does it take to actually write a book? The answer varies for every book and every author, but to get a more detailed answer be sure to read *Birth Of A Book* in the appendix section of *How To Write, Publish & Sell Your Own Aviation Books*.)

CHAPTER SIX

EDITING YOUR MANUSCRIPT

Editing is a necessary part of writing. Ernest K. Gann summed it up best in his autobiography, *A Hostage To Fortune*, when he wrote, "Never regard any of the words, sentences or even pages, with the fondness a father might feel for a newborn child. Cut mercilessly toward the clear and simple."[31]

Because I considered Gann's work to be the best in the aviation fiction genre, I tried to follow his advice the entire time I was writing *Black Falcon*. It paid off in the end when a book review in the February 1989 issue of *Waterfront News* stated, "If you like aviation stories that remind you more than a little of Ernie Gann, you'll like *Black Falcon*."[32]

Needless to say, I was totally thrilled that my book was compared to Gann's novels. How was it possible for me to achieve such a high compliment at the very beginning of my writing career? The answer lies in the editing stage of *Black Falcon*.

Actually, *Black Falcon* went through numerous editing stages. The first time was when I wrote the second draft of the manuscript. This editing stage was performed without the use of an outside editor. I realized early on that many of the characterizations in my initial draft were utterly horrible. Basically, the first draft was too bogged down with my own personal problems (which was one of the reasons I started writing the book in the first place), so I decided to begin the second draft by starting over from scratch. In fact, two of the characters from the first draft were completely removed from the second draft because they were getting in the way of the main story line.

After the second draft was completed, I performed a major editing job on it, mainly to achieve factual accuracy. I changed words, sentences and even pages - just like Gann had suggested in his autobiography. All authors should do this type of detailed editing, but many do not - hence, they give themselves away as being amateurs in a professional field.

A good example of editing for factual accuracy is the description I gave of a MAC-10 pistol in *Black Falcon*. Although this gun was manufactured as a legal semi-automatic weapon during the early 1980s, many smugglers were easily converting them into illegal automatic pistols. As an automatic, they became the weapon of choice among drug runners. Throughout my novel I referred to MAC-10s as automatic pistols, which in reality they were. After *Black Falcon* was published, however, one reader made a concerted effort to inform me that he was an expert on weapons and that it was obvious to him that I knew nothing about guns because my book had referred to MAC-10s as being automatic pistols rather than semi-automatics. I kindly explained to the man that drug runners usually used MAC-10s in their converted automatic modes. That little conversation taught me a valuable lesson on how nitpicky readers can be when it comes to authenticity.

I went through several more rewrites of *Black Falcon*, yet I still managed to miss a few inaccuracies that ended up in the book. One in particular was the spelling of Opa-locka Airport. During my research I had used a Miami Sectional Aeronautical Chart for plotting flight routes into and out of South Florida. On this government-issued chart, Opa-locka Airport was printed as *Opa Locka Airport*, so that's how I spelled it throughout my manuscript. I never took into account the real spelling of Opa-locka, which uses a hyphen between the two words and a lower-case *l* (as opposed to an upper-case *L*). What is so ironic about this mistake is that I had worked at Opa-locka Airport for three years (from 1979 to 1981) and never noticed it was spelled wrong in my manuscript.

Well, a magazine editor who reviewed *Black Falcon* in the Summer 1989 issue of *Airliners* caught my error and wrote, "For a local author, I was surprised to see Opa-locka incorrectly written."[33] Fortunately, the remainder of the book review was excellent, telling about my detailed descriptions of drug flights aboard radial-engined airfreighters. What did I learn from this review? I learned that book reviewers can be just as nitpicky about accuracy as ordinary readers. The only difference is that book reviewers can make or break a book in terms of sales.

Now I'm going to tell you how I actually edited my book. As you already know, I wrote the first and second drafts of *Black Falcon* by hand. When it came time for editing, I would read each chapter out loud to myself and make any necessary corrections to the handwritten pages as I went along. Figure 6-1 shows a page from the second draft of *Black Falcon* after the necessary changes had been written down, which made this the third draft of the manuscript.

At this point, I typed the manuscript on a small portable typewriter. I used plain white paper, double-spaced with wide margins. Then I found a quiet place and read through the entire manuscript at one sitting for the first time. I made changes as I went along, which became the fourth draft of *Black Falcon*. These changes were handwritten onto the typed pages. I then read through the manuscript again, making even more changes. I was now on the fifth draft of *Black Falcon*. But I still wasn't finished. I read through the manuscript one more time, made a few final changes, then

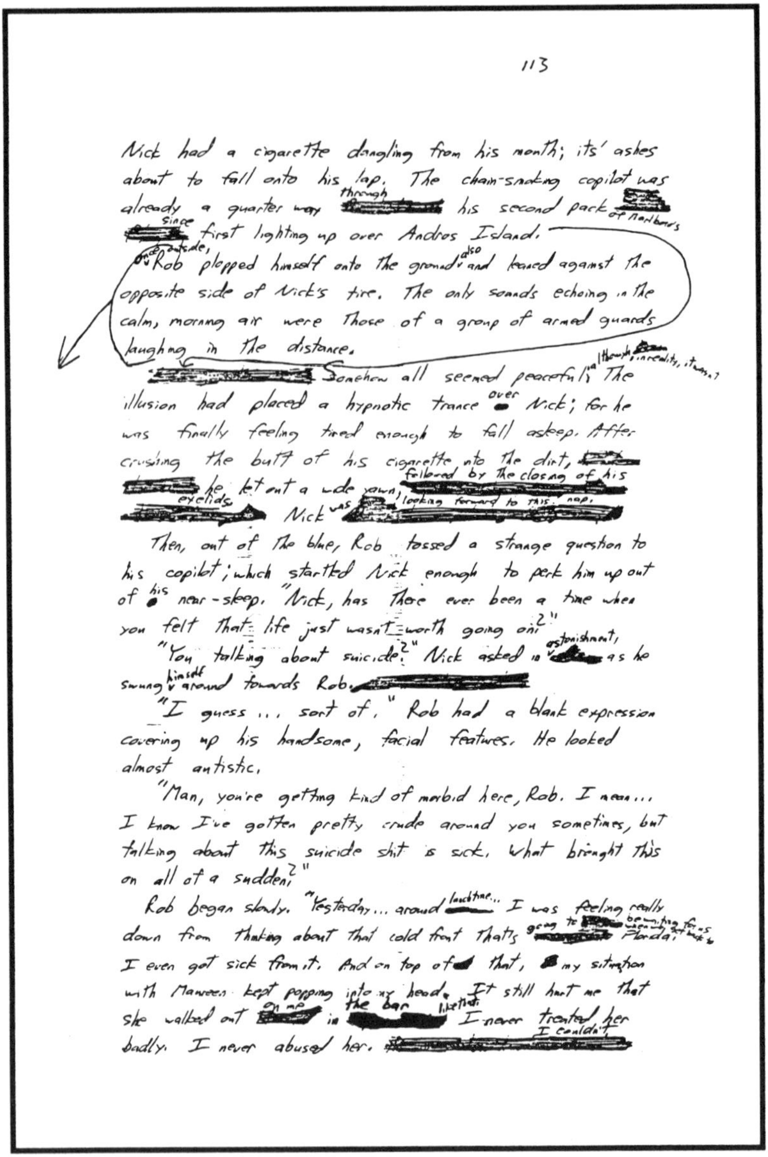

113

Nick had a cigarette dangling from his mouth; its' ashes about to fall onto his lap. The chain-smoking copilot was already a quarter way through his second pack of Marlboro's since first lighting up over Andros Island.

Once outside, Rob plopped himself onto the ground and also leaned against the opposite side of Nick's tire. The only sounds echoing in the calm, morning air were those of a group of armed guards laughing in the distance.

Somehow all seemed peaceful, although, in reality, it wasn't. The illusion had placed a hypnotic trance over Nick; for he was finally feeling tired enough to fall asleep. After crushing the butt of his cigarette into the dirt, he let out a wide yawn, followed by the closing of his eyelids. Nick was looking forward to this nap.

Then, out of the blue, Rob tossed a strange question to his copilot; which startled Nick enough to perk him up out of his near-sleep. "Nick, has there ever been a time when you felt that life just wasn't worth going on?"

"You talking about suicide?" Nick asked in astonishment, as he swung himself around towards Rob.

"I guess ... sort of." Rob had a blank expression covering up his handsome, facial features. He looked almost autistic.

"Man, you're getting kind of morbid here, Rob. I mean... I know I've gotten pretty crude around you sometimes, but talking about this suicide shit is sick. What brought this on all of a sudden?"

Rob began slowly. "Yesterday... around lunchtime... I was feeling really down from thinking about that cold front that's going to be waiting for us when we get back to Florida. I even got sick from it. And on top of that, my situation with Maureen kept popping into my head. It still hurt me that she walked out on me in the bar like that. I never treated her badly. I never abused her. I couldn't

Figure 6-1

re-typed the sixth draft of the manuscript. This was the draft that I sent to all of the major New York publishers.

I'm not sure if your laughing or crying at all the work I put myself through while editing *Black Falcon*, but I'm glad I suffered through it. Just keep in mind that when you are reading through your manuscript, if you stumble on any phrases, if certain words don't sound right, if something seems wrong or out of place, then change it. Keep doing this until your manuscript sounds *perfect*. I wrote seven drafts of *Black Falcon* before I was satisfied with the manuscript. *How To Get Published - Guaranteed* went through four drafts before it was finalized. *How To Get Published - Guaranteed*, by the way, was also written by hand and then typed using my portable electric typewriter. The book you are now reading, *How To Write, Publish & Sell Your Own Aviation Books*, was my first attempt at writing a manuscript using a computer, as opposed to handwriting. It only went through two drafts. Making editorial changes to the manuscript was much easier using the computer and I would recommend this method of writing to anyone thinking about starting a book. But if you're like I was (scared of computer technology), don't worry. I have proven that all a person really needs to be an author is a pen and lots of paper. It just takes a bit of extra work, that's all.

Did I say that *Black Falcon* went through seven drafts? That's right! You see, after all the New York publishing houses rejected the sixth draft, I decided to do one final draft before self-publishing my book - this time using an outside editor. You see, I had read my novel so many times that I had finally become immune to its problems, which included grammatical errors, some incorrect word usage, wrong phrases, a little bit of phony dialogue, and missing pieces of information. So I figured that an outside editor would be able to find any problems that might still be hidden within the text. In other words, a reading of the manuscript from a fresh perspective - a different point-of-view.

How did I choose my editor? Well, it had a lot to do with luck. You see, several family members had already read my manuscript and each of them said that *Black Falcon* was great. I knew, however, that *Black Falcon* was not great - just good. I knew it needed a bit more editorial work - a touch-up, so to speak. Then a woman who I worked with at the airport, Beth Hagan, asked if she could read the manuscript. I agreed to let her read it. After she finished going over *Black Falcon*, she came to me and said something like, "It's a good story, but *this* character can't possibly know *this* fact at *this* point in time." She then continued telling me where the story was lacking and where it was fine. I was not hurt by Beth's honesty. Instead, I was overjoyed to finally have gotten true constructive criticism regarding *Black Falcon*. Although Beth had never done any previous book editing, I asked her to be my editor for the seventh draft of the manuscript. She accepted my offer.

Beth and I went over my manuscript word-by-word, line-by-line, paragraph-by-paragraph, page-by-page, until *Black Falcon* was finally perfect. (Yes, I know I said it was perfect after the sixth draft, but I was wrong.) Many editorial changes were made during that two month process. Then, in the quiet of Beth's living room, she read the entire manuscript out loud to me. I kept my eyes closed and listened intently to the smoothness of the words, the rhythm of the story, the power of the various characters - my final reassurance that *Black Falcon* was ready to be published.

How much did it cost me to hire Beth as my editor? Well, she offered to do it for free - something fun to do - but I insisted on paying her for her editorial assistance. Because I was working on a tight budget, I told Beth that I would pay her $1,000 when *Black Falcon* came out in its second printing. In other words, if the book was a failure, no money would change hands. Well, a few years later I had the pleasure of presenting Beth with a $1,000 check from Tri-Pacer Press - signed and delivered during a celebration of the second printing of *Black Falcon*. That was the only time I ever enjoyed paying someone so much money.

How can *you* find editorial assistance for *your* book? A good place to start is the English Department at a local high school or college. First, introduce yourself as a local author and self-publisher, then explain to them what I did with *Black Falcon* and how I arranged my editing deal with Beth Hagan. Show them the cover of *How To Write, Publish & Sell Your Own Aviation Books* so they can see the quality of *Black Falcon* in the cover photograph. Offer some of the teachers the opportunity to edit your manuscript under the same terms I had offered to Beth. You'd be surprised at how many people would enjoy doing a project like this. I know this is true because I worked as a substitute teacher on a part-time basis at a local high school for six years and I got to know many of the English teachers.

Oh, don't forget to make copies of all your editorial changes and put them in the vault! This is very important. After all, you never know when disaster will strike.

Before ending this chapter, I'd like to explain two major changes that occurred during the editing stage of *Black Falcon*. Neither of these changes involved editorial corrections, but both of them contributed heavily to the eventual success of the published book.

The first major change concerned the title of the book, which I had originally titled just plain *Falcon*. I had picked *Falcon* (a deadly bird of prey) because of the novel's theme, which revolved around aerial drug smuggling. I wanted a title that evoked *death by force*. When my relatives started reading the manuscript, some of them thought the title was named after the drug kingpin character in the story, known by the code name Falcon. But this was not the case. In fact, the drug kingpin was written as a minor character in the story. The title *Falcon* was actually named after the gold falcon charms that the drug kingpin forced each of his underlings to wear around their necks as symbols of intense obedience and loyalty towards the organization. I had gotten the idea for these falcon charms after reading a newspaper article about the infamous Black Tuna Gang, an actual drug smuggling organization whose members wore black tuna charms around their necks. To solve my problem with the title, I knew it had to be changed. *Gold Falcon* didn't sound deadly enough, so I changed the color of the falcon charms from gold to black and retitled the novel *Black Falcon*. This turned out to be a very powerful title indeed and prevented any further confusion between the drug kingpin and the charms.

The second major change involved my use of a pseudonym instead of my birth name. Why did I choose not to use my real name as the author of *Black Falcon*? Well, for two reasons actually. First of all, using a pseudonym gave me the freedom to have two separate identities, which allowed me to create the illusion that Tri-Pacer Press was larger than a one-person publishing company. In other words, Chevy Alden would be known as the author of the action-adventure novel *Black Falcon*, while Richard Cook would be known as the editor-in-chief of Tri-Pacer Press. Anybody who is thinking about self-publishing should seriously consider using a pseudonym. If you want to use your real name as the author, then use a pseudonym as the editor-in-chief.

The second reason I used a pseudonym was so that I could create an image that was larger than life. Let's face it, Richard Cook is a rather ordinary name. It doesn't attract much attention to itself. Why do you think movie stars and musicians use pseudonyms? Do you think that Elton John would have gotten very far in the music industry if he had used his birth name, which was Reginald Kunneth Dwight? Many authors have also found success using pseudonyms.

For example, Samuel Clemens used the pseudonym Mark Twain. So I figured I'd use Chevy Alden to achieve my success. And guess what? It worked! And it's still working!

Why? Because the pseudonym Chevy Alden attracts attention to itself. It is highly unusual and it presents a perfectly balanced look when seen on any printed media (meaning that *Chevy* and *Alden* are the same length). It also works because Alden starts with an *A*. If I had chosen Voisin as the last part of my pseudonym, the *V* in Voisin would have worked against me. You see, listings are usually done in alphabetical order. As a struggling author who was seeking publicity, I knew I had to build as much name recognition as possible. That was done using a last name that began with the letter *A*. Figure 6-2 shows a portion of the program from the 3rd Annual Literary Forum & Author's Luncheon, which took place in Palm Beach County on May 25, 1989. I was one of the Honored Guests at that spectacular event. As you can see in Figure 6-2, I was given top-billing on the program (ahead of such famous authors as F. Lee Bailey,

Honored Guests

Chevy Alden	Edith Delatush	Nick Nixon
Adelaide Altman	Nicole Duplaix	Roxanne Pulitzer
F. Lee Bailey	Barbara Gibson	Patrick Smith
Joella Cain	James Hall	Dr. Barry Russal
Sydney Clary	C. Colburn Hardy	James Warnke
Linda Marx Cole	The Honorable James Knott	Bob Weller
Donald Curl	Sharon Komlos	Donald Wenberg
Sue Dauphin	Bruce Nash	

Figure 6-2

James Hall, Roxanne Pulitzer and Patrick Smith). This kind of top-billing has happened over and over again during my writing career and I'm sure it will continue in the future as long as I keep using my pseudonym (which I plan on doing).

How did I create my pseudonym? Well, at various times during my life I've owned a Chevy Chevelle, a Chevy Camaro and a Chevy Blazer. Need I say more in regards to the first part of my new identity? As for Alden - well, that part of my pseudonym came from John Alden, a classic sailboat designer during the early 1900s. I thought Alden sounded great with Chevy, so that's the name I picked back in 1987. It was as simple as that.

And now to sum up editing. A lot of authors find this process to be an activity that drains their creative energy. Others just don't like discarding anything they've already written. But editing is a necessary evil. Yes, it is draining - very much so! But it must be done. Always keep in mind the advice given by Ernest K. Gann in *A Hostage To Fortune*, which was to "Cut mercilessly toward the clear and simple."[34] That one statement sums up editing in its entirety.

CHAPTER SEVEN

HOW TO START YOUR OWN PUBLISHING COMPANY

Setting up a small publishing company in your home is not very difficult. All you need is a desk, a bookcase, a filing cabinet, a good typewriter (or personal computer), some basic office supplies, a finished manuscript to publish, and a few thousand dollars for book publishing expenses. You do *not* need a typesetting machine, a printing press, a book binding machine or an in-house art studio. These tasks will all be contracted out to professionals.

The first step in setting up your business is to pick a company name. It should be a name that means something personal to you. For example, I chose the name Tri-Pacer Press because it represents my dream of someday buying a Piper Tri-Pacer.

Here are a few other examples of how self-publishers named their publishing companies. Dan Poynter, the author of *The Self-Publishing Manual*, chose Para Publishing as a company name because his first book was about sport parachuting. Denton Rickey Moore, the author of *Gentlemen Never Sail To Weather*, chose the name Prospector Press because *Prospector* was the name of his sailboat. Are you getting a clear picture of how self-publishers pick names for their companies? The names are usually meaningful to the authors. When it comes time for you to choose a name for your publishing company, be sure it is one you'll be able to live with for a long time, because once your company is operational, you won't be able to change it (at least not easily).

Names of businesses are known as fictitious names. Most counties (and some states) have laws stating that fictitious names must be registered with the appropriate authorities. This is an easy process that anyone can do on their own. It involves getting a Fictitious Name Affidavit from your local county government. If your state also requires you to register a fictitious name, your county government will most likely advise you of this and possibly even provide you with the necessary forms.

Before sending in the completed Fictitious Name Affidavit, you will be required to place a classified advertisement in your local newspaper under the heading Notice Under Fictitious Name Law. This advertisement must run for a specified number of days, depending on the laws of the state and county in which you live. Your local newspaper will then give you a notarized statement saying that the advertisement was placed in the paper according to local and state laws.

Figure 7-1 shows the notarized statement I received from *The Sun-Tattler* stating that I ran a Notice Under Fictitious Name Law advertisement according to the laws of Broward County, Florida, and that this advertisement ran on April 13, 20, 27 and May 4, 1988. (Although I came up with the idea to self-publish my novel in 1987, I spent several months rewriting the seventh draft of *Black Falcon*, so I didn't actually get around to registering Tri-Pacer Press until 1988.) In the upper right-hand corner of Figure 7-1 you can see a copy of the actual advertisement, which *The Sun-Tattler* attached to their notarized statement.

Once you receive your notarized statement, you can then file your Fictitious Name Affidavit with your local county (and your state, if required). Figure 7-2 shows the Fictitious Name Affidavit I filed with Broward County,

THE SUN-TATTLER

88397962

Established as The Hollywood Sun · January 4, 1935
HOLLYWOOD, BROWARD COUNTY, FLORIDA

NOTICE UNDER FICTITIOUS NAME LAW
Notice is hereby given that the undersigned, RICHARD COOK, desiring to engage in business under the fictitious name, TRI-PACER PRESS, CHEVY ALDEN, intends to register said name with the County Administrator, Department of Finance, Division of Recording of Broward County, Florida.
/s/ Richard Cook
S-T April 13, 20, 27, May 4, 1988 (B-65)

STATE OF FLORIDA
COUNTY OF BROWARD
Before the undersigned authority personally appeared Jack D. Pate
who on oath says (he/she) is Advertising Sales Manager
of THE SUN-TATTLER, a daily newspaper published at Hollywood in Broward County, Florida; that the attached copy of advertisement, being a
Notice Under Fictitious Name Law
in the matter of Tri-Pacer Press , Chevy Alden
in the County Administrator xCourtx
was published in said newspaper in the issue of
April 13, 20, 27 & May 4, 1988

Affiant further says that the said SUN-TATTLER is a newpaper published at Hollywood in said Broward County, Florida, and that the said newspaper has heretofore been continuously published in said Broward County, Florida, each week and has been entered as second class matter at the post office in Hollywood in said County, Florida, for a period of one year next preceding the first publication of the attached copy of advertisement; and affiant further says that he has neither paid nor promised any person, firm or corporation any discount, rebate, commission or refund for the purpose of securing this advertisement for publication in the said newspaper.

Jack D. Pate

Sworn to and subscribed before me
this day of May A.D. 1988
Bob O. Burge
NOTARY PUBLIC
(SEAL)

Figure 7-1

FICTITIOUS NAME AFFIDAVIT

(Registration under Sec. 865.09, Fla. Statutes)

STATE OF FLORIDA)
)
COUNTY OF BROWARD)

Before me, the undersigned authority, an officer duly authorized to administer oaths and take acknowledgments, this day personally appeared the undersigned affiant, who being by me first duly sworn, depose(s) and say(s) that the names of all those interested in the business enterprise are:

Richard Cook

That the extent of the interest of each of them is:

100%

That the fictitious name under which said business is carried on is:

TRI-PACER PRESS, CHEVY ALDEN

That the location of said business is:

[illegible]

Mailing Address: P.O. Box 840111, Pembroke Pines, FL 33084

(Must be signed by all parties having an interest in this fictitious name.)

Signature
Address
Signature
Address
Signature
Address

"OFFICIAL NOTARY SEAL"
MARILYN V. SEMORA
MY COMM. EXP. 8/18/91

RECORDED IN THE OFFICIAL RECORDS BOOK OF BROWARD COUNTY, FLORIDA
L. A. HESTER
COUNTY ADMINISTRATOR

Sworn to and subscribed before me in the State and County aforesaid this 3rd
day of October, 1988

Marilyn V. Semora
Notary Public.

Figure 7-2

Florida. (Notice that I filed two fictitious names - one for Tri-Pacer Press and one for Chevy Alden.)

At that time, fictitious names were not required to be registered with the state of Florida, but a law has since been passed that requires fictitious names to be registered every five years.

The above mentioned forms are nothing more than paperwork. You don't need a lawyer to do this paperwork. Just get the proper forms from the appropriate government offices, enter the information and turn in the forms. There's usually a fee associated with each step of this process, but these fees are usually not more than $100 in most counties and states.

Your next step is to go to your local post office to rent a post office box for your business. This will be the mailing address for your publishing company. For instance, my business address is: Tri-Pacer Press, P.O. Box 840111, Pembroke Pines, Florida, 33084-2111. A small annual fee will be charged for the use of this post office box.

Next, find a good local bank and open up a business account, known as a DBA account. My account is listed as Richard Cook DBA Tri-Pacer Press, which stands for Richard Cook doing business as Tri-Pacer Press. This allows me to deposit checks that are made out to Tri-Pacer Press. In order to get a business account, you will once again have to show proof that you have filed a Fictitious Name Affidavit. Business checks will then be issued under your company's name. Be sure to have your business address printed on your checks.

Your next step will be to get an Occupational License from the county in which you live (and possibly from the city in which you live, depending on local laws). Again, this is simply a matter of filling out the appropriate forms and paying a small annual fee. For example, my Occupational License from Broward County, Florida, cost me $45 when I first filled out the application and has cost me an additional $45 each year since.

Figure 7-3 shows the Occupational License Applicant's Preliminary Information Sheet I had to fill out before receiving my Broward County Occupational License as a publisher. (Notice that I used publisher, rather than author, for my occupation. You should do the same.) Figure 7-4 shows a copy of the actual license, which is renewed annually.

All businesses that sell merchandise must collect state and local sales taxes, which then must be forwarded to the appropriate state and local revenue collection agencies. Before I was allowed to legally collect taxes, I had to file an application with the Florida Department of Revenue to get a Certificate of Registration/Sales Tax Certificate Number.

Figure 7-5 shows a copy of my Certificate of Registration/Sales Tax Certificate Number. The purpose of this document is two-fold. First, because I produce books for resale, I do not have to pay sales tax on the cost of producing the books. In other words, I am tax-exempt for this portion of my business. Secondly, this document gives me the authority to collect tax from retail customers when they purchase books from Tri-Pacer Press. I then send these taxes to the Florida Department of Revenue on a quarterly basis. For example, if a book sells for $18.95 at full retail and the state sales tax is 6%, then I have to collect $1.13 in tax every time I sell a copy of that book to a retail customer. Every three months I tally up all the sales taxes I have collected for that quarter and send a check to the state for that amount. This is not a difficult task. It simply requires basic record-keeping and collection skills.

If you feel uncomfortable handling any of the business procedures I have mentioned so far in this chapter, free help is available to you from your local chapter of SCORE (Service Corps of Retired Executives). This helpful organization is made up of men and women who have spent their lives in

PLEASE ANSWER ALL QUESTIONS

OCCUPATIONAL LICENSE APPLICANT'S PRELIMINARY INFORMATION SHEET

(FILL OUT ONLY WHEN APPLYING FOR A NEW LICENSE)

APPLICANT: PLEASE COMPLETE THIS FORM SO THAT WE MAY EXPEDITE THE PROCESSING OF YOUR BROWARD COUNTY OCCUPATIONAL LICENSE.

1. DATE BUSINESS OPENED October 1, 1988 OR DATE BUSINESS WILL OPEN

2. IS YOUR BUSINESS WITHIN THE JURISDICTION OF A CITY IN BROWARD COUNTY? YES: X NO:___ IF NO, DO YOU HAVE A ZONING CERTIFICATE? YES:___ NO:___ COPY MUST BE PRESENTED.

3. BUSINESS LOCATION: [redacted]
Street City Zip Code

STORE:___ SUITE NO:___
OFFICE:___ BAY NO:___
WAREHOUSE:___ EXPLAIN:___
HOME: X
OTHER:___

4. BUSINESS PHONE:___ HOME PHONE:___

5. NAME OF BUSINESS: TRI-PACER PRESS
SOLE PROPRIETOR X PARTNERSHIP ___ CORPORATION___ PA___
NAME OF OWNER, PRINCIPAL OR OFFICER: Richard Cook

6. TYPE OF BUSINESS: Publisher of Adventure Novels

7. MAILING ADDRESS, IF DIFFERENT FROM #3: P.O. Box 840111, Pembroke Pines, 33084
Street City Zip Code

8. NUMBER OF EMPLOYEES (INCLUDING OWNER AND PRINCIPALS) 1

9. CONTRACTORS MUST PRESENT CERTIFICATE OF COMPENTENCY CARD OR STATE CERTIFICATION CARD.

10. HAVE YOU BEEN ISSUED A NOTICE OF VIOLATION FROM AN INSPECTOR? YES:___ NO: X IF YES DATE OF VIOLATION___

11. HAVE YOU HAD A BROWARD COUNTY OCCUPATIONAL LICENSE BEFORE? YES:___ NO: X IF YES TYPE OF BUSINESS:___

12. DO YOU PRESENTLY HAVE OTHER LOCATIONS IN BROWARD COUNTY? YES:___ NO: X LOCATION(S)___

13. ARE THERE ANY COIN-OPERATED MERCHANDISE, SERVICE OR AMUSEMENT MACHINES ON THE PREMISES? YES:___ NO: X

DATE: 10/3/88

NAME OF APPLICANT Richard Cook
(PLEASE PRINT)

SIGNATURE OF APPLICANT [signature]

TITLE Editor-in-Chief, Owner

FOR OFFICE USE ONLY: VALIDATION___ ACCOUNT NO.___

FORM NO. 401-279a

Figure 7-3

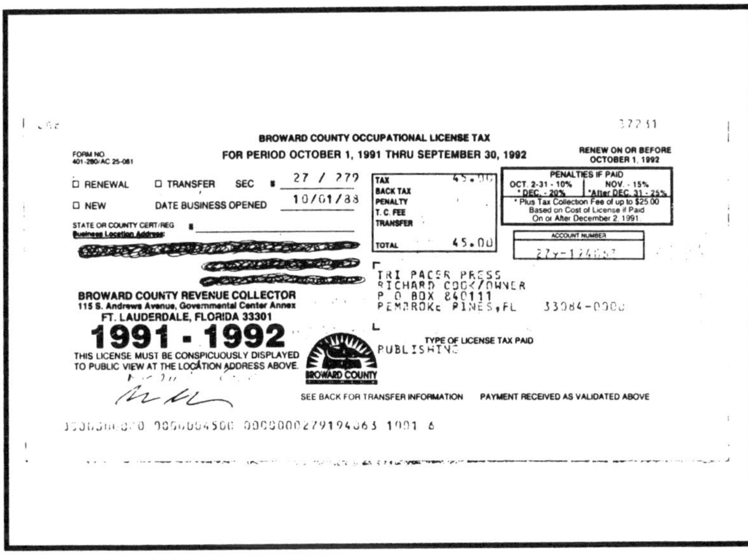

BROWARD COUNTY OCCUPATIONAL LICENSE TAX

FORM NO. 401-280/AC 25-081

FOR PERIOD OCTOBER 1, 1991 THRU SEPTEMBER 30, 1992

RENEW ON OR BEFORE OCTOBER 1, 1992

☐ RENEWAL ☐ TRANSFER SEC # 27 / 279

☐ NEW DATE BUSINESS OPENED 10/01/88

STATE OR COUNTY CERT/REG #

Business Location Address:

TAX	45.00
BACK TAX	
PENALTY	
T. C. FEE	
TRANSFER	
TOTAL	45.00

PENALTIES IF PAID
OCT. 2-31 - 10% NOV. - 15%
*DEC. - 20% *After DEC. 31 - 25%
* Plus Tax Collection Fee of up to $25.00 Based on Cost of License if Paid On or After December 2, 1991

ACCOUNT NUMBER

TRI PACER PRESS
RICHARD COOK/OWNER
P O BOX 840111
PEMBROKE PINES, FL 33084-0000

BROWARD COUNTY REVENUE COLLECTOR
115 S. Andrews Avenue, Governmental Center Annex
FT. LAUDERDALE, FLORIDA 33301

1991 - 1992

THIS LICENSE MUST BE CONSPICUOUSLY DISPLAYED TO PUBLIC VIEW AT THE LOCATION ADDRESS ABOVE.

TYPE OF LICENSE TAX PAID
PUBLISHING

BROWARD COUNTY

SEE BACK FOR TRANSFER INFORMATION PAYMENT RECEIVED AS VALIDATED ABOVE

Figure 7-4

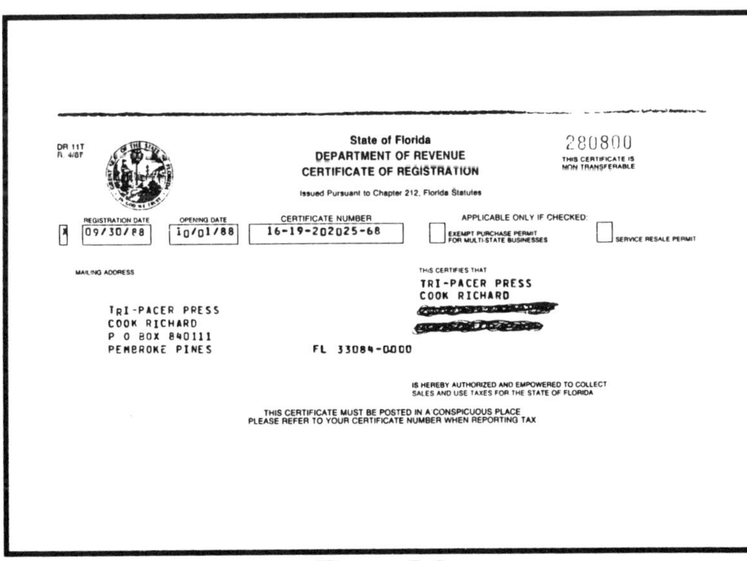

DR 11T R. 4/87

State of Florida
DEPARTMENT OF REVENUE
CERTIFICATE OF REGISTRATION

280800
THIS CERTIFICATE IS NON TRANSFERABLE

Issued Pursuant to Chapter 212, Florida Statutes

REGISTRATION DATE	OPENING DATE	CERTIFICATE NUMBER	APPLICABLE ONLY IF CHECKED:
09/30/88	10/01/88	16-19-202025-68	☐ EXEMPT PURCHASE PERMIT FOR MULTI-STATE BUSINESSES ☐ SERVICE RESALE PERMIT

MAILING ADDRESS

TRI-PACER PRESS
COOK RICHARD
P O BOX 840111
PEMBROKE PINES FL 33084-0000

THIS CERTIFIES THAT
TRI-PACER PRESS
COOK RICHARD

IS HEREBY AUTHORIZED AND EMPOWERED TO COLLECT SALES AND USE TAXES FOR THE STATE OF FLORIDA

THIS CERTIFICATE MUST BE POSTED IN A CONSPICUOUS PLACE
PLEASE REFER TO YOUR CERTIFICATE NUMBER WHEN REPORTING TAX

Figure 7-5

the business world and who have dedicated their retirements to help ordinary people start small businesses all across the country. SCORE can provide you with free business counseling, brochures and seminars on how to set up and run a small business. I attended a free seminar given by the Broward County Chapter and learned how to set up Tri-Pacer Press. Up until that time, I knew nothing about running my own business. After all, I'd been an aircraft refueler since the age of seventeen.

I also purchased an 80-page book (self-published by the Broward County Chapter of SCORE) titled *Small Business Manual - Basics Of Running A Small Business*. A nominal $5 fee was charged for this book, which included chapters on the following subjects: Business Plan, Marketing, Sales, Advertising, Finance, Insurance, Bookkeeping, Laws & Licenses, Library Facilities & Sources Of Information, Glossary Of Business Terms, and Elements Of Success & Failure.

One process which SCORE could not help me with was getting Tri-Pacer Press entered into the International Standard Book Numbering (ISBN) system and obtaining an ISBN Publishers Prefix and an ISBN Log Book. The ISBN system is an international data recording system for keeping track of published books. Publishers throughout the world use the ISBN system to list the titles of their books, prices, ordering information, discount schedules, names of authors and lots of other useful data. Libraries and bookstores use this system to obtain current information for ordering books that are still in print. In order for your publishing company to be considered a *real* publishing company, you must apply for an ISBN Publishers Prefix.

To enter your publishing company into the ISBN system, write to R.R. Bowker for an Application For ISBN Publishers Prefix. R.R. Bowker is the United States Agency for the ISBN system. Their address is: ISBN U.S. Agency,

R.R. Bowker, 121 Chanlon Road, New Providence, NJ 07974. When you receive the appropriate application, fill it out completely and mail it back to R.R. Bowker. There was no charge for this service when I originally applied back in 1987, but by the early 1990s a one-time service fee of $115 was being charged for first-time applicants. This fee might be slightly higher by now, due to the number of people starting their own publishing companies. Figure 7-6 shows the completed application I sent in for Tri-Pacer Press.

R.R. Bowker will process your application and assign an ISBN Publishers Prefix to your publishing company, along with an ISBN Log Book containing 10, 100 or 1,000 individual computer-generated ISBN Numbers, depending on the number of books you plan to publish. (Larger publishers can request ISBN Log Books that contain up to a million individual ISBN Numbers.) They might also assign you to a local R.R. Bowker Data Collection Center for future contacts. For example, I have been assigned to their Data Collection Center in Tampa, Florida.

To better understand how this ISBN system works, refer to Figure 7-7, which shows the ISBN Log Book that was assigned to Tri-Pacer Press. Notice that each of the individual numbers has ten digits that are hyphenated into four separate parts. Each part gives coded information. Let's use the ISBN Number 0-9619991-5-2 as an example since this is the ISBN Number assigned to the hardcover edition of *Black Falcon*. The first part of this number is the language code of the country where the book was published. The number *0* indicates that the book was published in an English-speaking country. The second part of this ISBN Number is the ISBN Publishers Prefix. The prefix *9619991* is the code for Tri-Pacer Press. No other publishing company can use this number. The third part of this particular ISBN Number is the suffix assigned to a specific edition of a specific book. The suffix *5* is the code I assigned to the hard-

INTERNATIONAL STANDARD BOOK NUMBERING. UNITED STATES AGENCY
International Standard Numbering System for Books, Software, Mixed Media etc.
in Publishing, Distribution and Library Practices
245 West 17th Street, New York, New York 10011 212-337-6971

International Standard ISO 2108

R.R. Bowker • The Information Company • A Division of Reed Publishing U.S.A.

December 9, 1987

RE: Letter of 11/25

Chevy Alden
TRI-PACER PRESS
P O Bx 840111
Pembroke Pines, FL 33084

APPLICATION FOR ISBN PUBLISHERS PREFIX: Printed Books & Pamphlets, Mixed Media Publications, other similar media including Educational Films/Video & Transparencies, Books on Cassettes, Microcomputer Software, Electronic Publications.

ISBN PREFIX IS ASSIGNED TO THE PUBLISHER (NOT PRINTER)

In order to assign the ISBN PUBLISHERS PREFIX, we require the following: (Print or Type)

WHAT IS YOUR PUBLISHING COMPANY NAME? Tri-Pacer Press
ADDRESS: P.O Box 840111 CITY: Pembroke Pines ST. FL ZIP: 33084
IF P.O. BOX, YOUR STREET ADDRESS IS REQUIRED [illegible]

1. Year you started publishing 1988
2. Number of titles published in 1986 .... ∅
3. Number of titles published in 1987 .... ∅
4. Number of titles planned for 1988 1
5. Number of titles you plan to do yearly ..... not more than 1/yr.
6. Number of titles still available (backlist). ∅
7. Are you a Sub./Div. of another company? No Give name, address and ISBN Prefix:

Have novel outlines being worked on

YOUR CATALOG AND PROMOTION MATERIALS ARE REQUIRED

8. Do you have other publishing subsidiaries/imprints in the U.S. or foreign country? No
   Give name, address, ISBN Prefix:
9. Do you distribute for any U.S./foreign publisher? No Give name (s), address (es) and ISBN Prefix:
10. Are you distributed or represented by another company in the U.S. or foreign country? No
    Give name (s), address (es)

President/Director: Richard Cook-Editor-in-Chief Date: 12/14/87
Name your Company ISBN Coordinator: Richard Cook-Editor-in-Chief Phone:

RETURN THIS FORM FOR THE ASSIGNMENT OF YOUR ISBN PUBLISHER PREFIX.

Emery Koltay
Emery Koltay, Director
ISBN/SAN Agency

PS: When our form is returned, we will assign an ISBN Publisher Prefix and send you a block of ISBN. Please note our new address above. Thank you.

Figure 7-6

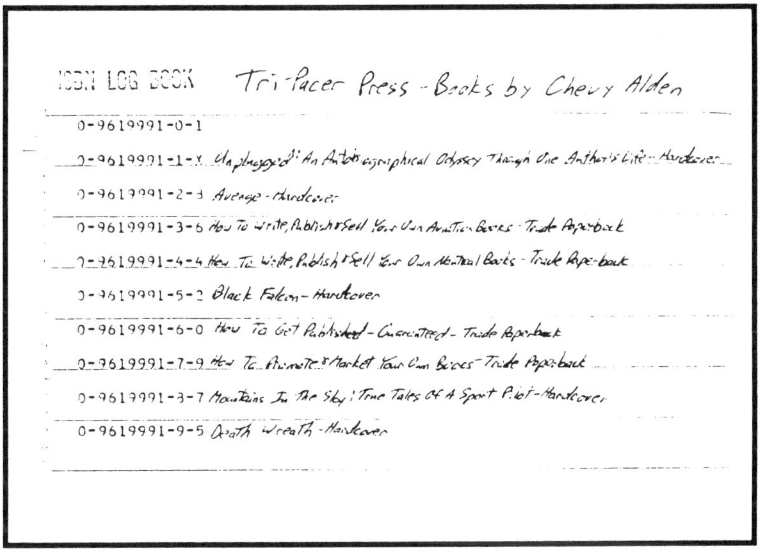

ISBN LOG BOOK Tri-Pacer Press - Books by Chevy Alden

0-9619991-0-1

0-9619991-1-X Unplugged: An Autobiographical Odyssey Through One Author's Life - Hardcover

0-9619991-2-8 Avenge - Hardcover

0-9619991-3-6 How To Write, Publish & Sell Your Own Aviation Books - Trade Paperback

0-9619991-4-4 How To Write, Publish & Sell Your Own Nautical Books - Trade Paperback

0-9619991-5-2 Black Falcon - Hardcover

0-9619991-6-0 How To Get Published - Guaranteed - Trade Paperback

0-9619991-7-9 How To Promote & Market Your Own Books - Trade Paperback

0-9619991-8-7 Mountains In The Sky: True Tales Of A Sport Pilot - Hardcover

0-9619991-9-5 Death Wreath - Hardcover

Figure 7-7

cover edition of *Black Falcon*. The fourth part of this ISBN Number is a computer-generated check digit that is used to make sure that no two ISBN Numbers will ever be duplicated accidentally. The check digit for the hardcover edition of *Black Falcon* is *2*, which has no real meaning other than being a computer-generated check digit.

An ISBN Log Book comes to publishers as a group of numbers. It is then up to publishers to assign one edition of each book to each individual ISBN Number. For example, if I were to publish a paperback edition of *Black Falcon*, I would have to assign a different ISBN Number to it than the one I use for the hardcover edition. Looking once again at Figure 7-7, you can see that I have already assigned most of my remaining ISBN Numbers to various titles, even though most of these books are not yet written (although they are all outlined). For instance, 0-9619991-9-5 is the ISBN Number assigned to the hardcover edition of *Death Wreath* (the first sequel to *Black Falcon*), while the ISBN Number 0-9619991-2-8 has been assigned to the hardcover edition of *Avenge* (the second sequel to *Black Falcon*).

Because each individual ISBN Number can only be assigned to one edition of one book, it seems that I will have a problem in the future when I start publishing paperback editions of my novels (or when I come up with ideas for additional books). In other words, I will run out of ISBN Numbers before I run out of new books to publish. The ISBN system recognizes this problem and there are two easy solutions to it. When a publisher runs out of ISBN Numbers, they can either reapply for an additional set of ISBN Numbers, which will give their publishing company a new ISBN Publishers Prefix (meaning that they will now have two active prefixes), or they can start an imprint publishing company (sort of a subsidiary to their original publishing company) and apply for a new ISBN Publishers Prefix along with a new set of ISBN Numbers for their imprint. Many of the larger publishing companies use the imprint method, but it's probably a lot easier for smaller publishers not to get involved with this, since it involves additional paperwork (such as Fictitious Name Affidavits and other stuff).

Putting my original example back together again, the ISBN Number 0-9619991-5-2 stands for the hardcover edition of *Black Falcon*, which was published by Tri-Pacer Press in an English-speaking country. When a library or bookstore gets a request to order the hardcover edition of *Black Falcon* (such as when a customer walks in off the street and makes a special request for this title), they most likely have no idea who the author or publisher is, or how to order a copy of the book. The acquisition person at the library or bookstore would look up *Black Falcon* in *Books In Print* (R.R. Bowker's publishing database) and find out that the publisher of the book is Tri-Pacer Press and that the author is Chevy Alden. They would also get the current price of the book and ordering information. Hopefully, they will then order the book.

Does the ISBN system work? Well, take a look at Figure 7-8, which shows an actual order I received in 1994 for one copy of *How To Get Published - Guaranteed* from a bookstore in Elking Park, Pennsylvania. I do not know anybody who lives in Elking Park, Pennsylvania, so this order request came from somebody outside my circle of family and friends. Simply by word-of-mouth (or perhaps because of a book review), someone heard about *How To Get Published - Guaranteed* and walked into their local bookstore to place an order for my book. The bookstore used *Books In Print* and sent an order to Tri-Pacer Press. I've gotten orders from all across the United States because of *Books In Print*. Many orders will come in according to ISBN Numbers rather than titles, so these numbers are extremely important.

How do publishers get their books listed in *Books In Print*? Easy! When you send in your Application For ISBN Publishers Prefix, be sure to request a free copy of *The ISBN System User's Manual*, which explains in detail how

BILL TO:
ELKINS PARK, PA 19117-2698 USA
OFFICE:
FAX:

SHIP TO:
ELKINS PARK, PA 19117-2698 USA

ORDER DATE: 4/01/94

TRI-PACER PRESS 2474417-
P O BOX 840111
PEMBROKE PINES, FL 33084-2111

SHIP A S A P WITHIN CONTINENTAL USA VIA UPS
PLEASE FURNISH 2 COPIES OF INVOICE

OUR P.O. NUMBER F4684 MUST APPEAR ON ALL LABELS, INVOICES AND PACKING SLIPS

QUANTITY ORD. SHPD	AUTHOR	TITLE	ISBN
1	ALDEN	HOW TO GET PUBLISHED - GUARANTEED : A SELF-HELP ORIG. P288 P. ILL. (PAPER IF AVAILABLE)	0961999160

SPECIAL INSTRUCTIONS: SHIP AVAILABLE TITLES ASAP. BACKORDER UNAVAILABLE TITLES. CONSOLIDATE BACKORDER SHIPMENTS.
HARD COVER WHEN AVAILABLE, UNLESS OTHERWISE SPECIFIED.
IF HARD COVER UNAVAILABLE, PROVIDE PAPERBACK. PROVIDE MAXIMUM DISCOUNT TO US.

PURCHASER'S SIGNATURE..........

Figure 7-8

the ISBN system works. (Don't you just love how I'm telling you how to obtain all this free information?) Also, ask for an Advanced Book Information form (known as an ABI form within the publishing industry), which will be used to get your book listed in *Forthcoming Books* and *Books In Print*. These important listings are absolutely free, so take advantage of the services provided by R.R. Bowker.

Approximately six months before your book is actually published, you should fill out the ABI form and send it in to R. R. Bowker. This will get your book listed in the next available issue of *Forthcoming Books* and every issue thereafter of *Books In Print* (which is updated annually). Both of these reference manuals (also published in CD-ROM format) are used extensively for book ordering purposes within the industry, so you definitely want to get listed in these directories.

Figure 7-9 shows the ABI form I sent to R.R. Bowker for *Black Falcon*. As you can see, it is a fairly simple form to complete. I typed the title of the book, the publisher's name and address, the publisher's phone number, the author's name, the page count of the book (which I listed as 224 pages, but which ended up being 257 pages), the primary audience (adult male), the type of book (fiction), the publication date (October 1988), the price of the book (which I listed as $16.95, but which was increased to $17.95 prior to the publication of the first printing), the ISBN Number (0-9619991-5-2), the Library Of Congress Catalog Card Number (88-50158 - which I will explain how to obtain in Chapter Nine), and a brief description of the book's subject matter.

Figure 7-10 shows the ABI form I sent to R.R. Bowker prior to the publication of *How To Write, Publish & Sell Your Own Aviation Books*. In comparing the ABI form in Figures 7-9 against the ABI form in Figure 7-10, you can see that this form changed slightly during the ten year pe-

ADVANCE BOOK INFORMATION

TITLE: BLACK FALCON

SUBTITLE:

SERIES:

Foreign Language: Translation ☐, from what language:

AUTHOR(S): Chuvy Alden

EDITOR(S):

TRANSLATOR(S):

ILLUSTRATOR(S):

INTRO. BY; PREFACE BY; etc.

ILLUSTRATIONS YES ☐ NO ☐

PAGES: Approx. 224

AUDIENCE (Select Primary Audience): Adult Male

College Text☐ Young Adult ☐ Grade:

Elhi Text ☐ Grade: Juvenile ☐ Grade:

Original Paperback ☐

Revised ☐ Abridged ☐ 2nd Ed. ☐ Other:

PUBLICATION DATE: October 1988

Reprint ☐ If reprint, name of orig. publisher & orig. pub. date:

ISBN NOTE: Put full 10 digit number in spaces below. The system requires a separate ISBN for each edition.

On short discount (20% or less) ☐

ENTER PRICE(S) BELOW:		INT'L STANDARD BOOK NUMBER
HARDCOVER TRADE	$16.95	ISBN 0-9619991-5-2
If juv., is binding guaranteed?		
LIBRARY BINDING		ISBN
HARDCOVER TEXT		ISBN
PAPER TRADE		ISBN
PAPER TEXT		ISBN
TCHRS. ED.		ISBN
WKBK.		ISBN
LAB MANUAL		ISBN
OTHER SPECIFY		ISBN

LC# 88-50158

Order # (optional):

PUBLISHER (Not Printer): TRI-PACER PRESS
Address: P.O. Box 840111
Pembroke Pines, FL 33084

Telephone (305) 432-9683

DISTRIBUTOR, if other than publisher:
(If you distribute foreign books you must be their exclusive U.S. distributor. Please send us a copy of your documentation for exclusivity)

IMPRINT (If other than company name):

THIS WORK IS ESSENTIALLY (Check one):

☒ FICTION ☐ TEXTBOOK
☐ POETRY ☐ BIOGRAPHY
☐ DRAMA ☐ OTHER ______ Specify
☐ CHILDREN'S FICTION
☐ ESSAYS

PRIMARY SUBJECT OF BOOK
(Be as specific as possible):

BLACK FALCON is an adventure novel revolving around the life of an unstable, South Florida drug pilot. When he quits the drug trade at the wrong time, he enters into a desperate struggle for survival against his distraught emotions, nature's elements and the deadly world of narcotics smuggling. This novel takes the reader from a sleazy Miami bar to a salty Key West tavern, from the cockpit of a drug plane to a clandestine airfield in the mountains of Colombia, from a tropical Bahamian cruise aboard a private sailboat to an action-packed ending on Ft. Lauderdale's Venice-like canal system. The novel's overall message is clearly anti-drug.

Completed by: Richard Cook
Editor-in-Chief

Figure 7-9

ADVANCE BOOK INFORMATION

TITLE: How To Write, Publish & Sell Your Own Aviation Books

SUBTITLE:

Title Volume Number:

PUBLISHER (Not Printer): Tri-Pacer Press

ADDRESS: P.O. Box 840111, Pembroke Pines, FL 33084-2111

Is this a set? Yes☐ No☐ Number of Volumes:

PHONE: 954-432-9683 TOLL FREE:

FAX: 305-652-6296 TELEX:

Is Title Orig. Paperbk? ☒YES ☐NO

E-mail, Internet, etc.:

Pub. Date: (MM/YY) 01/98

IMPRINT (if other than company name):

Copyright Date: (year) 1998

DISTRIBUTOR (if other than publisher):

(If you distribute foreign books, please send us a copy of your documentation and indicate whether distribution is exclusive or non-exclusive. Currently, only exclusive distributors are included in the US portion of the BIP database.)

Pages: 350 Illustrated?: ☒YES ☐NO

CONTRIBUTOR NAME (A=Author, E=Editor, I=Illustrator, P=Photographer, T=Translator):

Contributor Chevy Alden ☒A ☐E ☐I ☐P ☐T ☐Other (check all that apply)

Contributor Walter J. Boyne ☐A ☐E ☐I ☐P ☐T ☒Other Foreword (check all that apply)

Contributor Richard Cook ☐A ☒E ☐I ☐P ☐T ☐Other (check all that apply)

Contributor ☐A ☐E ☐I ☐P ☐T ☐Other (check all that apply)

Series Title:

Series Subtitle: Series Volume Number:

Edition Info.: ☐Reprint ☐Revised ☐Abridged ☐Large Type ☐Unabridged ☒Other(specify) 1st Edition

Edition No.: ☐2nd ☐3rd ☐4th ☐5th ☐6th ☐Other(specify)

Audience: (children's books require grade levels) GRADES: AGES:

☐Juvenile ☐Young Adult ☐ElHi—Reading Levels

☐College ☒Adult\General ☐Other (specify)

Original Title: (if previously published & changed)

Translated Title:

Current Language (if other than English): Original Language (if translated):

Publisher Order No.: LCCard No(LCCN):

Point Size (Large Type Bks.): Book Size: 5½" x 8½" Book Weight:

For ISBN Applicants: Fill out this form for each of your titles and return with your application. When the top copy is returned to you with your ISBN log book, assign an ISBN to each title, enter it below, and return the form(s) to the R.R. Bowker Data Collection Center, Florida. This will allow us to list your title(s) with all information in our database.
ISBN Note: Write full 10-digit number in space below. The BIP system requires a separate ISBN for each edition and binding.

BINDING	ISBN	PRICE	*TYPE:	**CURRENCY	DISCOUNT
Hardcover: ☐Trade ☐Textbk		$			Short ___% Max ___
Paperback: ☒Trade ☐Textbk	0-9619991-3-6	$21.95	Retail	US	Short 20 % Max 50
☐Library Binding		$			Short ___% Max ___
☐Mass Market		$			Short ___% Max ___
☐Other:		$			Short ___% Max ___
☐Other:		$			Short ___% Max ___

(OTHER refers to Tchr's. ed., Wkbk., Lab Manual, Looseleaf, etc.)

*Price type refers to invoice, retail, tentative, etc.
**Currency refers to US, Canadian, etc.

(Short Discount is 20% or less)

Type of Work: ☒Non-Fiction ☐Fiction ☐Poetry ☐Drama ☐Essay ☐Other

Subject Area: ☐Children's (CB) ☐Law(LB) ☐Med(MB) ☐Relig(RB) ☐Sci-Tech(ST) ☒Other Writing & Publishing

Description of Content: A self-help manual that teaches aviators how to write, publish and sell their own aviation books. Based on the author's success at writing, publishing and marketing his aviation thriller "Black Falcon". Dozens of photos and illustrations.

Figure 7-10

riod since *Black Falcon* was published. You can also see that some of the information provided on the ABI form is just a preliminary estimate (such as the page count), which is why this form is called an Advanced Book Information form. For example, I didn't yet have the Library Of Congress Catalog Card Number for *How To Write, Publish & Sell Your Own Aviation Books* when I sent in the ABI form for this book.

Just prior to the publication of *Black Falcon*, R.R. Bowker sent me a computerized checklist on which I had to make any changes or updates to my listing for the next issue of *Books In Print*. I wrote the final page count and the new price on the checklist and mailed it back to R.R. Bowker. Each year, just prior to the publication of a new issue of *Books In Print*, a checklist arrives in the mail, assuring me that my listing for *Black Falcon* (as well as for my other books) will always remain up-to-date. Figure 7-11 shows the 1997 checklist for *Books In Print*.

R.R. Bowker also sends a second checklist each year so

RUN DATE: 2/04/97
SORT: PUB/IMPRINT/TITLE
R.R. BOWKER
P.O. BOX 6000-0103
OLDSMAR, FL. 34677-0103

TRI-PACER PR
REED REFERENCE PUBLISHING - BOWKER
BOOKS IN PRINT CHECKLIST

PAGE: 1
DEADLINE: 03/28/97
BOWKER CONTACT NAME: BIP STAFF
BOWKER CONTACT PHONE: 813-855-4635

WINTER 1997 MAILING

TITLE:Black Falcon
ISBN:0-9619991-5-2 CONTRIBS:AU Chevy Alden BINDING:Trade Cloth PRICE:~~$18.95~~ $20.95 PUB-DATE:10/1988 PGS:257 LCCN:88-050158
PUB-SYM:Tri-Pacer Pr

TITLE:How to Get Published - Guaranteed SUB-TTL:A Self-Help Manual for Assuring the Publication of Your Books ORG-PB:X
ISBN:0-9619991-6-0 CONTRIBS:AU Chevy Alden; ED Richard Cook BINDING:Trade Paper PRICE:$18.95 PUB-DATE:10/1994 PGS:~~220~~ 328
ILLUS:yes ~~AUDIENCE:C~~ Delete LCCN:93-060906 PUB-SYM:Tri-Pacer Pr

TOTAL NUMBER OF TITLES: 2
TOTAL NUMBER OF ISBNS: 2

Figure 7-11

that publishers can update information about their company. This checklist can be seen in Figure 7-12, which shows the changes I made in 1997.

As a small publisher, you should become a member of the National Association of Independent Publishers (NAIP). Through the *NAIP/Publisher's Report* newsletter, you will receive industry-related information, networking opportunities, listings of literary events and major book shows, numerous how-to articles, and other helpful advice for small publishers. Membership information for NAIP can be obtained by writing to: National Association of Independent Publishers, P.O. Box 430, Highland City, FL 33846-0430.

You should also join a regional publishing organization for networking purposes, similar to the Florida Publishers Association (the group I belong to). The following groups are the ones I know about: Arizona Book Publishing Association, Baltimore Publishers Association, Bay Area Independent Publishers Association, Book Publishers Northwest, Book Publishers of Texas Association, Colorado In-

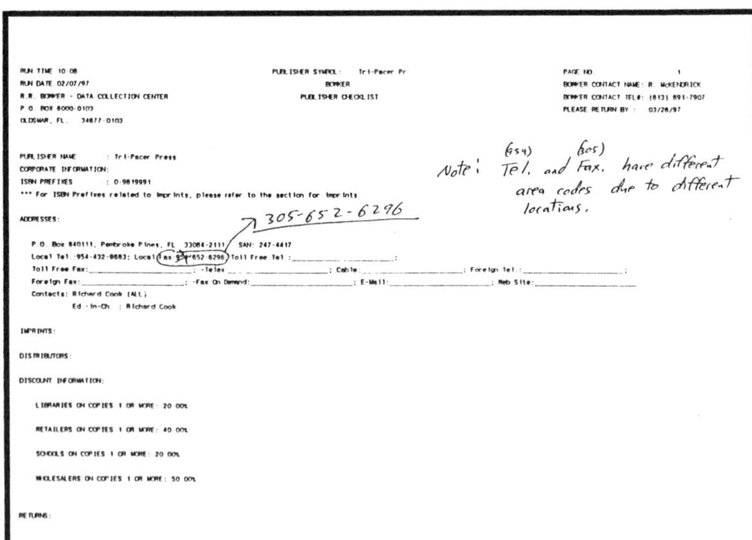

RUN TIME 10:08
RUN DATE 02/07/97
R.R. BOWKER - DATA COLLECTION CENTER
P.O. BOX 6000-0103
OLDSMAR, FL. 34677-0103

PUBLISHER SYMBOL: Tri-Pacer Pr
BOWKER
PUBLISHER CHECKLIST

PAGE NO. 1
BOWKER CONTACT NAME: R. McKENDRICK
BOWKER CONTACT TEL#: (813) 891-7907
PLEASE RETURN BY: 03/28/97

PUBLISHER NAME : Tri-Pacer Press
CORPORATE INFORMATION:
ISBN PREFIXES : 0-9619991
*** For ISBN Prefixes related to Imprints, please refer to the section for Imprints

ADDRESSES:

P.O. Box 840111, Pembroke Pines, FL 33084-2111 SAN: 247-4417
Local Tel.:954-432-9683; Local Fax:954-652-6296; Toll Free Tel:_____;
Toll Free Fax:_____; -Telex_____; Cable:_____; Foreign Tel.:_____;
Foreign Fax:_____; -Fax On Demand:_____; E-Mail:_____; Web Site:_____
Contacts: Richard Cook (ALL)
Ed.-In-Ch. : Richard Cook

IMPRINTS:

DISTRIBUTORS:

DISCOUNT INFORMATION:

LIBRARIES ON COPIES 1 OR MORE: 20.00%

RETAILERS ON COPIES 1 OR MORE: 40.00%

SCHOOLS ON COPIES 1 OR MORE: 20.00%

WHOLESALERS ON COPIES 1 OR MORE: 50.00%

RETURNS:

Figure 7-12

dependent Publishers Association, Consortium of Northern Publishers, Florida Publishers Association, Independent Publishers Association of Canada, Mid-America Publishers Association, New England Publishers Association, Northern Book Publishers Consortium, Northwest Association of Book Publishers, Publishers Association of the South, Sacramento Publishers Association, San Diego Publishers Alliance, Tucson Book Publishers Association. There are possibly other regional organizations that I do not know about.

The next step in starting your own publishing company is to organize an office in your home. Choose a place that can be dedicated solely to your publishing needs - whether it be a spare bedroom, a den or even a portion of your living room if you cannot spare an entire room for your company. Figure 7-13 shows a photo of my office during the writing of *How To Get Published - Guaranteed* (before I bought a computer and modernized my business), when furnishings included an old wooden desk, filing cabinet, typewriter, two lamps, and several bookshelves filled with research material,

Figure 7-13

reference books, resource guides and directories. Once I finished *How To Get Published - Guaranteed*, my office was updated with a new spacious desk, Packard Bell computer and Hewlett Packard DeskJet printer. Figure 7-14 shows my office as it looks today, including the beautiful framed poster from one of my Barnes & Noble lectures (an accomplishment I'm extremely proud of after all these years, because it means I'm finally breaking into the world of big-time bookselling).

The portion of my office not shown in Figures 7-13 and 7-14 (the opposite side of the room) is used as a warehouse for my inventory of books. It is also used as a stock room for my shipping supplies, display boards, research material and other odds and ends related to the publishing business. Basically, my office is clean and simple.

The office supplies that I keep in my desk drawers include pencils, pens, tape, paper clips, business envelopes, company letterhead, printing paper for the computer, staples and stapler. The typewriter that I used to use was a good

Figure 7-14

quality Smith-Corona. I still keep this hanging around the office, because every once in a while I'll have to fill in the blanks on a form and it is easier to do it on the typewriter than trying to do it with the computer.

In my filing cabinet I have dozens of file folders for organizing paperwork and documents. The headings on these file folders are: Articles Of Interest, Articles Pending, Articles Published, Artist Contracts, Bank Statements, Bar Codes, Books In Print Updates, Book Vendors, Catalogs, Classified Ads, Copyright Information, Correspondence, Federal Income Tax Records, Fictitious Name Information, Florida Sales Tax Records, ISBN Information, Letters Of Interest, Library Of Congress Information, Marketing Plans, Master Copies, Occupational Licenses, Permission Grants, Printers, Purchase Orders, Receipts Current Year, Receipts Past Years, Reviews, Subsidiary Rights Sales, Typesetters. You will probably not need all of these file folders when you first set up your publishing company, but eventually you will have to add them to your filing system as your own business grows and prospers.

In order for your business to give off a professional appearance, you will need to print company stationery (known as letterhead). Your letterhead should include the name of your publishing company, the mailing address, city, state, zip code and telephone number. It can also include a company logo or slogan. The slogan on my letterhead during the first six years that Tri-Pacer Press was in existence read *Publisher of Quality Action-Adventure Novels*. Obviously, this slogan became obsolete once *How To Get Published - Guaranteed* came out in print, so I had to think up a new company slogan. As of 1994, my letterhead has read *Publisher of Quality Titles Since 1988*. I also had to print up new letterhead in 1996 because the phone company changed my telephone area code from 305 to 954.

All correspondence should be on high-quality paper and

should be printed using either an electronic typewriter with a correctable film ribbon or a high-quality printer (inkjet or laser). Manual typewriters and dot-matrix printers do not produce professional results.

One final note about organizing your business. Be sure to make photocopies of everything you mail out (letters, applications, contracts, manuscripts, business documents, tax statements, etc.) just in case you need to refer to a particular item a few years down the road.

If you can accomplish everything I've just described to you in this chapter, then you have the ability to start your own publishing company. If you get to this stage of the game, please let me know by sending me a note on your company's letterhead. Also, when your book comes out, please send me a complimentary review copy so I can add it to my growing collection of self-published aviation books.

Now that you know how to set up a small publishing company, you are ready to learn how to develop a realistic marketing plan for your book. Chapter Eight will teach you how to do this, but first take a short break to rest your eyes, stretch your legs and relax your mind.

CHAPTER EIGHT

DEVELOPING A MARKETING PLAN FOR YOUR BOOK

Most books are written for specific markets. In order for these books to reach their intended markets, they must be promoted in ways that will bring them to the attention of the people who will most likely want to read the books. This is where self-publishing blossoms as a viable way to market books - especially aviation books! You see, large publishers sell their books primarily through bookstores, hoping that their titles will be seen and purchased by the general public. In reality, however, marketing books in this manner assures that a good percentage of the books will be returned (unsold and oftentimes damaged) to the publishers, with the sure-fire bestsellers (books written by well-known mega-authors and celebrities) ending up earning most of the profits for the bookstores and publishers.

In fact, the average book put out by a major publisher spends only about one month in print.[35] When I first self-published *Black Falcon* in 1988, the average time a book spent on the bookstore shelves was three to six months, so times have gotten much worse for the major publishing industry. I mean, why even publish a book in the first place if it is only going to be available for a month?

A self-publisher, on the other hand, can get around this situation by developing a marketing plan that is geared specifically towards the readership of his or her book - a marketing plan that will assure continued sales year after year (as long as the book is well-written), slowly building up a customer base and total income from that book (which is what I've done with *Black Falcon* and what you can do with your aviation book).

What do I mean by a marketing plan? Simply that the promotion and sales effort for each self-published book should be aimed directly at the people for whom the book was written. It is unrealistic to think that a book will sell well just because it is placed on bookstore shelves. However, it *is* realistic to assume that the same book *will* sell well if it is placed in specialty stores that cater to the interests of the book's intended readership, as well as being displayed and sold at conventions and trade shows dealing with the book's specific subject matter. For example, the intended readership for *Black Falcon* when I first wrote the book were pilots and sailors (a double bonus for me as far as marketing was concerned). By stocking my air-sea novel at aviation and nautical supply stores, as well as displaying it at aviation and nautical trade shows, I have done quite well selling *Black Falcon*. In fact, it wasn't until *How To Get Published - Guaranteed* came out in print that the large chain bookstores even took notice of my books, although several dozen independent bookstores had been selling *Black Falcon* right from the start.

The key to a self-publisher's success is to avoid thinking in terms of how the major publishers do things - after all, many of them are losing money as a whole (even with their bestsellers bringing in millions of dollars). Selling self-published books is an entirely different ball game - which is great because the major publishers don't play this game, leaving the field wide open for opportunities. For example, one particular air show I did in 1996 cost me approximately $700 in expenses, but I sold $2,700 worth of books (well worth the effort and expense).

How does an aviation self-publisher develop a realistic

marketing plan for his or her book? Well, the first step involves figuring out exactly who the book was written for. In other words, who will be the primary audience? It is possible, of course, to have more than one primary audience. *Black Falcon* is a perfect example of this because it was written for pilots and sailors - two very different groups of readers. Another good example is *How To Get Published - Guaranteed*, which was written to appeal to all types of audiences, since people from all walks of life want to learn how to publish their own books.

A funny thing happened, though, when I started selling *How To Get Published - Guaranteed*. It seemed that pilots had more stories to tell than any other group of people - and they were buying *How To Get Published - Guaranteed* like crazy. But because the title wasn't aviation-oriented, FBOs, aviation bookstores, pilot shops and aviation catalogs just wouldn't stock this book, nor did I get very many book reviews from aviation magazines, although I did get a few that produced excellent mail-order sales. Then it occurred to me that since *Black Falcon* was an aviation-oriented book, and since *How To Get Published - Guaranteed* explained how I wrote, published and sold out the first printing of *Black Falcon*, and since pilots were buying my how-to book like crazy, then why not write a new version of *How To Get Published - Guaranteed* aimed specifically towards aviators and title the new edition *How To Write, Publish & Sell Your Own Aviation Books*? I thought this would surely solve my problem of getting my how-to book distributed within the aviation industry. And so, the book you are now reading was born!

But I still wasn't finished! To increase the likelihood of media attention and hopefully lots of sales, I came up with the idea of finding a famous aviation author to write a foreword to this book. I had no idea how to accomplish such a grand task (especially since I was not very well known as a writer), so I figured I'd just start at the top and ask. I definitely wanted to find an author whose aviation books I had read, someone who was well respected within the entire aviation industry, someone who was a pilot, and someone who believed in the worth of self-publishing. That was a lot to ask for, but I came up with just such a person in Walter J. Boyne, the former director of the National Air and Space Museum (NASM/Smithsonian Institution). If you haven't already read Boyne's foreword at the beginning of this book, do it right now. The foreword, along with Boyne's name on the cover of *How To Write, Publish & Sell Your Own Aviation Books*, has surely helped me accomplish my goal of selling how-to books within the aviation industry.

Every publisher knows that a well-written foreword by a well-chosen author can add greatly to a book's promotional potential. *Well-written* and *well-chosen* are the key words here! Boyne's foreword was definitely well-written. As far as being well-chosen, a list of Boyne's many accomplishments surely proves that he was the *write person* (pun intended) to write the foreword to my book. Boyne is a bestselling author of both fiction and nonfiction, the founder of *Air & Space/Smithsonian* magazine, and a retired colonel from the United States Air Force with more than 5,000 hours in a variety of aircraft, including the B-47 and B-52. A partial book list (the ones that I have in my personal collection) of his more than two dozen published books includes *Trophy For Eagles*, *Eagles At War*, *The Wild Blue* (a bestseller co-authored with Stephen L. Thompson), *The Leading Edge*, *The Smithsonian Book Of Flight*, *The Smithsonian Book Of Flight For Young People* (which I bought for my daughter) and *Silver Wings: A History Of The United States Air Force* (my personal favorite).

Getting back to my original point here, you must figure out who the primary audience for your book will be. To do this, you must figure out who will be most interested in the particular subject matter of your book. This primary audience, no matter who they are, will be the target of your initial marketing plan.

Later on, after you have gained some experience and have built up a sales history and publishing track record, you can expand into other markets, constantly building a base for future sales. John Kremer's book *1001 Ways To Market Your Books* is an excellent resource manual for finding new ways to increase your sales. This book has become a trusted business tool for many self-publishers (myself included). It's like having a personal marketing consultant at home on a 24-hour-per-day basis. (To order a copy of *1001 Ways To Market Your Books*, please refer to the *Writer's Book Mart Catalog* in the appendix section of *How To Write, Publish & Sell Your Own Aviation Books*.)

When I first thought about self-publishing *Black Falcon*, my initial marketing plan was quite simple. I figured I could probably sell my air-sea adventure novel at aviation bookstores, nautical bookstores, flight shops, marine supply outlets, air shows and boat shows, plus get book reviews by sending press releases and review copies of *Black Falcon* to aviation and sailing magazines.

During the first two years as a self-publisher, I stuck closely to my initial marketing plan, doing very well in the process. During my third year, however, things really took off for me as I became better known within South Florida's literary community. Once that happened, I was able to expand into new markets, selling copies of my novel to book wholesalers, book distributors, libraries and independent bookstores, as well as to the general public (people who were neither pilots nor sailors). It was during my third year as a self-publisher that I also became heavily involved with doing book shows - an area of endeavor that has brought many new opportunities my way.

As you start coming up with new ideas for marketing

your book, write them down in a spiral notebook. This notebook will be your marketing plan - an important business tool that will be used again and again throughout your writing and publishing career. Be original with your marketing ideas! Be creative!

No matter what type of book you have written (or are planning to write), there will most likely be a certain portion of the population interested in reading it (whether they be female pilots, male pilots, military pilots, airline pilots, sport pilots, or just plain armchair pilots). You must figure out who these people are, where they gather, what associations and clubs they belong to, where they shop, what trade shows and conventions they attend, and which magazines they read. This is the trick to developing a realistic marketing plan. The one thing that makes this so easy is that *you* are the author of *your* book - therefore, *you* should already have firsthand knowledge about your book's subject area, as well as the answers to most of the above questions.

One of the most important aspects of your marketing plan will be the press releases and review copies you will be sending to magazines. Why? Because people tend to buy books that have been approved by knowledgeable sources (book reviewers). The time to start thinking about book reviews is during the developmental stage of your initial marketing plan. No matter what subject you have written about (or are planning to write about), there are magazines geared towards your subject. You should start early preparing a list of who will get press releases and review copies. Your list should include the names and addresses of the magazines, as well as the names of the editors.

A quick stop at your local newsstand will provide you with a small selection of periodicals (the industry term for magazines and newspapers), but don't stop there. Go to the reference section of any good library and ask to see either *The Standard Periodical Directory* or *Ulrich's Internation-*

al Periodicals Directory. Both of these excellent reference sources list the names and addresses of thousands of periodicals on every imaginable subject.

The Standard Periodical Directory (published by Oxbridge Communications) is considered to be the most comprehensive directory of United States and Canadian periodicals, with more that 75,000 listings (including consumer magazines, newsletters, newspapers and trade journals, plus ethnic, literary, museum, religious and social group publications). This directory is divided into 265 major subject categories, which are then subdivided into thousands of minor categories. I used this reference source extensively while developing the initial marketing plan for *Black Falcon*. What did I find? Well, I found more than 400 periodicals listed under *Aeronautics* and nearly 200 periodicals listed under *Boating*. From these selections, I chose approximately eighty magazines and newspapers that I felt would be interested in reviewing my novel.

Ulrich's International Periodicals Directory (published by R.R. Bowker) is the other reference source for finding listings of magazines and newspapers. I did not use this directory for *Black Falcon*, however, because I did not know about it at the time. *Ulrich's* lists more than 118,500 periodicals from throughout the world, all of which are divided into approximately 700 subject categories.

Getting book reviews is just one of the ways to bring attention to self-published books. Another way is through trade shows, conventions and other special events geared towards specific interests. Almost every subject has its own inner circle of associations, clubs and organizations. Most of these groups hold some type of meeting either monthly, bimonthly, semi-annually or annually. Research your particular subject area and make a list of the groups you'll likely want to contact once your book is published. With *Black Falcon*, for example, I've participated at air shows and aviation trade shows as both a book dealer and a lecturer. You should try to do the same.

Many self-publishers make the mistake of publishing their books first, then trying to figure out how to market the books. This wastes valuable time, so don't make this mistake yourself. A marketing plan should *always* be developed prior to a book's publication date. For example, before I even finished writing my second book, *How To Get Published - Guaranteed*, I already knew how I was going to market it, which basically came down to renting booth space at book shows, making guest appearances at book fairs and other literary events, and sending press releases and review copies to writing magazines. The initial marketing plan I had developed for *How To Get Published - Guaranteed* was the exact plan I used when the book first came out in print. It was simple, yet effective. Why was it effective? Because it cut out all wasteful sales efforts by narrowing down my readership into one specific group - authors. The fact that the book also appealed to pilots was a total surprise to me - a surprise that eventually led to the book you are now reading!

The purpose of this chapter has been to show you how to develop a realistic marketing plan for your book. Chapter Fifteen will show you how to implement your marketing plan, as well as how to expand upon your plan using a process known as networking. What is networking? Basically, it is the process of making personal contacts at one event, which will lead to other events, which will then lead to new personal contacts and even more events, and so on and so forth. Everything builds upon itself. This is how I sold out the first printing of *Black Falcon* without ever having sold a single copy of my novel through one of the major chain bookstores - positive proof that self-publishing works.

Tri-Pacer Press has been successful from the very beginning! By constantly updating my marketing plan and by continually pouring my earnings back into my publishing business (rather than wasting the money on frivolous expenditures), I have been able to build Tri-Pacer Press into a company that will someday (hopefully) allow me to quit my full-time job at the airport. In the meantime, I keep plugging away at my writing, publishing, bookselling, flying and sailing. Not a bad way to wander through life!

CHAPTER NINE

LEGAL CONSIDERATIONS

I am not an attorney - therefore, I cannot give you legal advice within the pages of this chapter. A qualified legal representative specializing in literary affairs should be consulted regarding all legal matters pertaining to your book. As in my own case, I have used the services of an attorney on two separate occasions during the course of my writing career. In each instance, the legal fee came to less than $200.

What I *can* offer you in this chapter is an overview of the legal matters I was up against while writing and self-publishing *Black Falcon*. Clearly understand that I am not giving you legal advice, but merely expressing factual accounts of my dealings with the legalities associated with the publishing industry. For me, these legalities came under four categories: Copyright Protection, Library Of Congress Catalog Card Number, Permission Grants, and Fair Use.

Copyright Protection. Under the United States Copyright Law, a piece of writing is copyrighted the moment the work is created in fixed form, meaning that an author does not legally have to register the work with the Copyright Office. On the other hand, it is a very good idea to go the extra step - hence, most authors *do* register their writing.

Back in 1981, when the total sum of my writing consisted of a 30-page expanded outline for the novel *Falcon* (later changed to *Black Falcon*), I was unaware of how the copyright registration process worked. To protect myself, I hired an attorney who specialized in Copyright Law. For a $110 fee, this attorney registered *Falcon* with the United States Copyright Office. The fee broke down into two parts - $10 for the registration fee and $100 for the attorney's fee. A few months later I received my Certificate Of Copyright Registration in the mail. Up until that time, I had always been worried that someone would steal my manuscript (a sure sign that I was still an amateur), but after I received my certificate the worry went away.

Since that early period of my writing career, I have learned that there is no need to hire an attorney for filing a copyright application. The procedure is very simple and merely involves filling out the proper form and sending it to the Copyright Office with a check for $20 (due to an inflationary price increase) and two copies of the best edition of the work being copyrighted once it is published or one photocopy of the work being copyrighted if it is not going to be published until a future date. That's what I did in 1986 when I copyrighted the completed manuscript for *Falcon*, then again in 1988 when I copyrighted the revised manuscript for *Black Falcon*.

To get the proper application forms, I sent a letter to the Copyright Office asking them to send me a free copy of Circular R1 (Copyright Basics), Circular R99 (Highlights Of The Current Copyright Law), Circular R96 (Methods Of Affirmation And Positions Of The Copyright Notice On Various Types Of Works) and several copies of Form TX (Application For Copyright Registration). The address of the Copyright Office is: Register Of Copyrights, Copyright Office, Library Of Congress, Washington, D.C. 20559.

When I received my packet of information from the Copyright Office in 1986, I discovered that since I was registering an unpublished manuscript, I had to type the following copyright notice on the first page of the manuscript: Unpublished Work (c) 1986 by Richard Cook. (Note: The *c* is suppose to have a circle around it, but I haven't figured out how to do this on my computer yet, which is why I used parentheses here.)

After filling out Form TX, I attached a check to it (made payable to the Register Of Copyrights), sealed it in a padded shipping envelope with a photocopy of the entire *Falcon* manuscript, then sent the entire package via certified mail to the address listed above. A few months later I received my Certificate Of Copyright Registration for *Falcon*. It was at that point that I started submitting *Falcon* to the major New York publishing houses, all of whom turned it down.

By the end of 1987 I had learned about self-publishing and decided to start Tri-Pacer Press. Because my manuscript was now going to be published and because it went through a major rewrite (as well as a title change), a new Form TX had to be sent in. The copyright notice also had to be changed to now read: Copyright (c) 1986, 1988 by Richard Cook. I was also required to send in two hardcover editions of *Black Falcon* once the book came out in print.

Figures 9-1 and 9-2 show the Form TX application that I sent in for *Black Falcon* in 1988. The instructions for using this form are attached to each application, so I will just go over them briefly here.

Section 1 shows the new title and the previous title of my novel.

Section 2 lists the author of the work. It reads: Richard Cook whose pseudonym is Chevy Alden. In other words, *Black Falcon* is copyrighted under my birth name even though it is written and published under my pseudonym. The reason for this is that a Certificate Of Copyright Registration is a tangible asset and can be inherited by heirs upon an author's death. A copyright that is filed under an author's birth name is good for the author's life span plus an additional fifty years, whereas a copyright filed under an author's pseudonym is only good for a period of seventy-five years from the book's original date of publication. Because *Black Falcon* was published when I was thirty years old, it was advantageous to my heirs to have the book copyrighted under my birth name. Also, it was a way for me to have my real name appear in the book (on the copyright page) as the author. Continuing on with Section 2, you will see that it also lists my birth year, nationality, and the nature of authorship, which for me included the entire 257-page novel, plus the jacket copy, the jacket art (which I did not create, but which was created for me under contract with the copyright signed over to me) and the jacket photograph (which was taken by my wife with the copyright signed over to me).

Section 3 shows the year the manuscript was completed and the date it was first published.

Section 4 lists the copyright claimant for *Black Falcon* and the claimant's address. I am the copyright claimant for *Black Falcon*.

Section 5 answers a few questions pertaining to the previous copyright registration for *Falcon*. Basically, it states that a previous registration had been filed with the Copyright Office and that a new registration was being filed because the work was now being published. Section 5 also lists the previous registration number, which was TXu 258-009 (the *u* meaning that it was previously unpublished) and the year of original registration. Notice in the top right corner of Figure 9-1 that the new registration number is TX 2-489-342 and that the *u* is gone (meaning that the work is now published).

Section 6 was not applicable to *Black Falcon*.

Section 7 lists the names and locations of the manufacturers who performed certain processes during the publication of my novel. (For full protection under the United States Copyright Law, it is a requirement in certain instances that the books must be manufactured in the United

CERTIFICATE OF COPYRIGHT REGISTRATION

UNITED STATES COPYRIGHT OFFICE · THE LIBRARY OF CONGRESS

OFFICIAL SEAL

This certificate, issued under the seal of the Copyright Office in accordance with the provisions of section 410(a) of title 17, United States Code, attests that copyright registration has been made for the work identified below. The information in this certificate has been made a part of the Copyright Office records.

REGISTER OF COPYRIGHTS
United States of America

FORM TX
UNITED STATES COPYRIGHT OFFICE

REGISTRATION NUMBER

TX 2 489 342

TX TXU

EFFECTIVE DATE OF REGISTRATION

11 14 88
Month Day Year

DO NOT WRITE ABOVE THIS LINE. IF YOU NEED MORE SPACE, USE A SEPARATE CONTINUATION SHEET.

1 TITLE OF THIS WORK ▼

Black Falcon

PREVIOUS OR ALTERNATIVE TITLES ▼

Falcon

PUBLICATION AS A CONTRIBUTION If this work was published as a contribution to a periodical, serial, or collection, give information about the collective work in which the contribution appeared. Title of Collective Work ▼

If published in a periodical or serial give: Volume ▼ Number ▼ Issue Date ▼ On Pages ▼

2 NAME OF AUTHOR ▼

a Richard Cook whose pseudonym is Chevy Alden

DATES OF BIRTH AND DEATH Year Born ▼ 1958 Year Died ▼

Was this contribution to the work a "work made for hire"? ☐ Yes ☒ No

AUTHOR'S NATIONALITY OR DOMICILE Name of Country United States OR { Citizen of ▶ Domiciled in ▶

WAS THIS AUTHOR'S CONTRIBUTION TO THE WORK Anonymous? ☐ Yes ☒ No Pseudonymous? ☒ Yes ☐ No — If the answer to either of these questions is "Yes," see detailed instructions.

NATURE OF AUTHORSHIP Briefly describe nature of the material created by this author in which copyright is claimed. ▼

Entire 257 page novel, jacket copy, jacket art, jacket photograph.

NOTE Under the law, the "author" of a ...

NAME OF AUTHOR ▼ b

DATES OF BIRTH AND DEATH Year Born ▼ Year Died ▼

Was this contribution to the work a "work made for hire"? ☐ Yes ☐ No

AUTHOR'S NATIONALITY OR DOMICILE Name of country OR { Citizen of ▶ Domiciled in ▶

WAS THIS AUTHOR'S CONTRIBUTION TO THE WORK Anonymous? ☐ Yes ☐ No Pseudonymous? ☐ Yes ☐ No — If the answer to either of these questions is "Yes," see detailed instructions.

NATURE OF AUTHORSHIP Briefly describe nature of the material created by this author in which copyright is claimed. ▼

02345010B

NAME OF AUTHOR ▼ c

DATES OF BIRTH AND DEATH Year Born ▼ Year Died ▼

Was this contribution to the work a "work made for hire"? ☐ Yes ☐ No

AUTHOR'S NATIONALITY OR DOMICILE Name of Country OR { Citizen of ▶ Domiciled in ▶

WAS THIS AUTHOR'S CONTRIBUTION TO THE WORK Anonymous? ☐ Yes ☐ No Pseudonymous? ☐ Yes ☐ No — If the answer to either of these questions is "Yes," see detailed instructions.

NATURE OF AUTHORSHIP Briefly describe nature of the material created by this author in which copyright is claimed. ▼

3 YEAR IN WHICH CREATION OF THIS WORK WAS COMPLETED This information must be given in all cases. 1988 ◀ Year

DATE AND NATION OF FIRST PUBLICATION OF THIS PARTICULAR WORK Complete this information ONLY if this work has been published. Month ▶ October Day ▶ 31 Year ▶ 1988 United States of America ◀ Nation

4 COPYRIGHT CLAIMANT(S) Name and address must be given even if the claimant is the same as the author given in space 2. ▼

Richard Cook

See instructions before completing this space

TRANSFER If the claimant(s) named here in space 4 are different from the author(s) named in space 2, give a brief statement of how the claimant(s) obtained ownership of the copyright. ▼

DO NOT WRITE HERE — OFFICE USE ONLY

APPLICATION RECEIVED NOV. 14. 1988

ONE DEPOSIT RECEIVED

TWO DEPOSITS RECEIVED NOV. 14 1988

REMITTANCE NUMBER AND DATE

MORE ON BACK ▶ • Complete all applicable spaces (numbers 5-11) on the reverse side of this page. • See detailed instructions. • Sign the form at line 10.

DO NOT WRITE HERE

Figure 9-1

EXAMINED BY

CHECKED BY

FORM TX

TX 2 489 342

☐ CORRESPONDENCE Yes

☐ DEPOSIT ACCOUNT FUNDS USED

FOR COPYRIGHT OFFICE USE ONLY

DO NOT WRITE ABOVE THIS LINE. IF YOU NEED MORE SPACE, USE A SEPARATE CONTINUATION SHEET.

PREVIOUS REGISTRATION Has registration for this work, or for an earlier version of this work, already been made in the Copyright Office? 5

☒ Yes ☐ No If your answer is "Yes," why is another registration being sought? (Check appropriate box) ▼

☒ This is the first published edition of a work previously registered in unpublished form.

☐ This is the first application submitted by this author as copyright claimant.

☐ This is a changed version of the work, as shown by space 6 on this application.

If your answer is "Yes," give: **Previous Registration Number** ▼ TXu 258-009 **Year of Registration** ▼ 1986

DERIVATIVE WORK OR COMPILATION Complete both space 6a & 6b for a derivative work; complete only 6b for a compilation. 6

a. **Preexisting Material** Identify any preexisting work or works that this work is based on or incorporates. ▼

b. **Material Added to This Work** Give a brief, general statement of the material that has been added to this work and in which copyright is claimed. ▼

See instructions before completing this space

MANUFACTURERS AND LOCATIONS If this is a published work consisting preponderantly of nondramatic literary material in English, the law may require that the copies be manufactured in the United States or Canada for full protection. If so, the names of the manufacturers who performed certain processes, and the places where these processes were performed must be given. See instructions for details. 7

Names of Manufacturers ▼	Places of Manufacture ▼
BookCrafters, Inc.	Chelsea, Michigan, U.S.A.
Tri-Pacer Press	Pembroke Pines, Florida, U.S.A.

REPRODUCTION FOR USE OF BLIND OR PHYSICALLY HANDICAPPED INDIVIDUALS A signature on this form at space 10, and a check in one of the boxes here in space 8, constitutes a non-exclusive grant of permission to the Library of Congress to reproduce and distribute solely for the blind and physically handicapped and under the conditions and limitations prescribed by the regulations of the Copyright Office: (1) copies of the work identified in space 1 of this application in Braille (or similar tactile symbols); or (2) phonorecords embodying a fixation of a reading of that work; or (3) both. 8

a ☐ Copies and Phonorecords b ☐ Copies Only c ☐ Phonorecords Only

See instructions

DEPOSIT ACCOUNT If the registration fee is to be charged to a Deposit Account established in the Copyright Office, give name and number of Account. 9

Name ▼ Account Number ▼

CORRESPONDENCE Give name and address to which correspondence about this application should be sent. Name/Address/Apt/City/State/Zip ▼

Richard Cook, [redacted]

Area Code & Telephone Number ▶ (305)432-9683

Be sure to give your daytime phone ◀ number

CERTIFICATION* I, the undersigned, hereby certify that I am the 10

Check one ▶
☒ author
☐ other copyright claimant
☐ owner of exclusive right(s)
☐ authorized agent of ______

Name of author or other copyright claimant, or owner of exclusive right(s) ▲

of the work identified in this application and that the statements made by me in this application are correct to the best of my knowledge.

Typed or printed name and date ▼ If this is a published work, this date must be the same as or later than the date of publication given in space 3.

Richard Cook date ▶ 11/6/88

Handwritten signature (X) ▼

MAIL CERTIFICATE TO 11

Name ▼ Richard Cook

Number/Street/Apartment Number ▼ [redacted]

City/State/ZIP ▼ [redacted]

Certificate will be mailed in window envelope

Have you:

- Completed all necessary spaces?
- Signed your application in space 10?
- Enclosed check or money order for $10 payable to *Register of Copyrights*?
- Enclosed your deposit material with the application and fee?

MAIL TO: Register of Copyrights, Library of Congress, Washington, D.C. 20559

* 17 U.S.C. § 506(e): Any person who knowingly makes a false representation of a material fact in the application for copyright registration provided for by section 409, or in any written statement filed in connection with the application, shall be fined not more than $2,500.

U.S. GOVERNMENT PRINTING OFFICE: 1987—181-531/40,024

April 1987 - 200,000

Figure 9-2

States or Canada.) Section 7 lists the manufacturers as BookCrafters (the book printer) in Chelsea, Michigan, and Tri-Pacer Press (the publisher) in Pembroke Pines, Florida (both being located in the United States of America).

Section 8 was not applicable to *Black Falcon*.

Section 9 lists the name, address and telephone number for all correspondence regarding the Form TX application.

Section 10 lists me as the person who filled out the application, along with my signature and the date the application was filled out.

Section 11 lists the address where the Certificate Of Copyright Registration would be mailed to.

That is all that is involved in registering a piece of writing with the United States Copyright Office. The easy-to-understand instructions that come with each Form TX application make it hard to make a mistake.

Library Of Congress Catalog Card Number. A Library Of Congress Catalog Card Number is a unique identification number assigned to a book by the Cataloging In Publication Division of the United States Library Of Congress. This number is used by libraries to locate information from national databases and also to order catalog cards from the Library Of Congress. It is possible to have this number preassigned to a book prior to the book's publication date by sending a Request For Preassignment Of Library Of Congress Catalog Card Number application to the Library of Congress. The address is: Library Of Congress, Cataloging In Publication Division, Washington, D.C., 20540. This is the same address used when ordering blank applications, as well as ordering a free instructional booklet titled *PCN Guidelines For The Preassigned Card Number Program*, which is available to anybody who asks for it.

The reason for obtaining a Catalog Card Number (LC# in the publishing industry) in advance of publication is so that it can be typeset onto the copyright page so that it appears in the book. I sent in my application several months prior to the actual publication of *Black Falcon*. Along with the application, I also had to send in a rough draft of the title page. Figure 9-3 shows the Request For Preassignment Of Library Of Congress Catalog Card Number application that I sent in for *Black Falcon*. Figure 9-4 shows the rough draft of the title page.

Because the Request For Preassignment Of Library Of Congress Catalog Card Number application is so simple and self-explanatory, I will not describe each step here like I did for the Form TX copyright application. However, please look at Step 9 in Figure 9-3, which asks for the approximate number of pages. Step 9 on my application shows 224 as the number of pages. In actuality, the final page count came to 257 pages. This is no big deal, so don't worry if it happens to you, too.

As you can see, applying for a Library Of Congress Catalog Card Number is just as easy as applying for a Certificate Of Copyright Registration - in fact, it's easier. It's also free. Just remember that when your book finally comes out in print, you must send one copy to the Library Of Congress for cataloging purposes. (Note: This copy is in addition to the two copies that must sent to the Copyright Office, but it is mailed separately.)

Permission Grants. When I originally wrote my novel, I wanted to use a portion of the lyrics from the following three songs: *A Pirate Looks At Forty* by Jimmy Buffett, *I Am - I Said* by Neil Diamond, and *Desparado* by Don Henley and Glenn Frey of the Eagles. I knew that I needed permission grants from the copyright holders of these songs in order to use the lyrics, otherwise I would have committed copyright infringement.

To seek permission grants for the above mentioned songs, I wrote letters to ASCAP and BMI, the two music industry clearinghouses that deal with song and lyric rights.

REQUEST FOR PREASSIGNMENT OF LIBRARY OF CONGRESS CATALOG CARD NUMBER

NOTE: Card numbers cannot be preassigned to books which are already published.

DATE: 2/3/88

...BLISHER'S NAME ON TITLE PAGE: TRI-PACER PRESS

YOUR NAME: Richard Cook PHONE NUMBER: ______

Type or print clearly the complete address to which the preassigned card number should be sent. (This will be your return mailing label.)

Richard Cook, Editor-in-Chief
TRI-PACER PRESS
P.O. Box 840111
Pembroke Pines, FL 33084

FOR CIP OFFICE USE
Library of Congress Catalog Card Number preassigned is:

Transcribe the information in items 1-8 exactly in the form and order in which it will appear on the title or copyright pages of the printed book. Use only those abbreviations which will actually appear on these pages. (**Please attach a copy of the proposed title page, if available.**)

1. Author(s) Chevy Alden
2. Editor(s) Richard Cook
3. Title BLACK FALCON
4. Subtitle ______
5. Edition (exactly as printed in the publication, e.g. second edition, revised edition, etc.) ______
6. U.S. place of publication: City Pembroke Pines State Florida
7. Any copublisher(s) and place ______
8. Series title and numbering, exactly as printed in the publication ______
9. Approximate number of pages 224
10. Number of volumes 1
11. ISBN 0-9619991-5-2
12. Binding: [X] Hardcover [ ] Paperback
13. Proposed date of publication: Month October Year 1988
14. Does (or will) the title in item 3 appear at periodic intervals, e.g. annually, quarterly, etc.? [ ] Yes [X] No
15. Will this publication be submitted later for CIP data? [ ] Yes [X] No
16. Language of text, if other than English ______

For each title which is preassigned a Library of Congress catalog card number, the Library of Congress requires one non-returnable complimentary copy of the best edition of the published book. If selected for the Library's collections, the book will be cataloged. A postage-free, self-addressed label will be sent with the preassigned card number for your convenience in mailing the required advance copy of the work as soon as printed. This copy is in addition to copyright deposit copies.

Send this form to: **Library of Congress**
Cataloging in Publication Division
Washington, DC 20540

FOR CIP OFFICE USE ONLY. DO NOT WRITE BELOW THIS LINE

Searching notes:

RECD: ______
ASGN: ______
SENT: ______
APIF: ______

607-7 (rev 10/86)

Figure 9-3

BLACK FALCON

CHEVY ALDEN

TRI-PACER PRESS

Figure 9-4

Well, either I didn't go about it in a professional manner or I didn't take the initiative far enough (or perhaps I just went about the whole process in the wrong way), but in any case, I came up empty-handed as far as permission grants were concerned. Because of this, I knew some rewriting was in store for me.

It was no big deal taking out the lyrics from *A Pirate Looks At Forty* and *I Am - I Said*, because they were only used in minor ways. Such was not the case, however, with *Desparado*. You see, several of the lines from this song were the basis for the emotional turning point in the love-hate relationship between two of my novel's main characters. The dilemma I was faced with was to make reference to the song without using any of the lyrics. After some deep sole-searching and creative ingenuity, I was finally able to rewrite the scene. Figure 9-5 shows the page from *Black Falcon* on which this scene appears. Read it and see how I solved my problem.

Besides lyrics, authors must also obtain permission for the use of poems, quotes from magazine articles, quotes from newspaper articles, and quotes from other books. In certain instances, however, writers can use portions of copyrighted material (without getting permission grants) under an axiom known as fair use. This principle can lead to some very touchy situations - therefore, writers must be extremely careful not to cross over the boundary separating fair use from unfair use.

Fair Use. The best explanation I have found for fair use was written by Bruce Henderson in his book *How To Bulletproof Your Manuscript* (published by Writer's Digest Books). He wrote, "Writers do have the protected privilege of using another's work - with proper credit given - in a *reasonable manner* without having to secure the permission of the author. This is called *fair use*, and *fair* is the key word here. Courts have held that such use promotes intellectual

88 BLACK FALCON

someday you'll be very sorry because you're going to wake up one cold morning and you're going to run your tired hands through your hair, which by then will be gray, and you're going to suddenly realize that you're an old man and that your whole life has somehow slipped by you. And when the lonely chill of the air clutches your body beneath the covers and you roll over to cuddle up to somebody warm, that somebody's not going to be there. Your bed will be empty. Not because your lover got up to get a drink of water or to use the bathroom. But because she was never there to begin with. Because you never let her in. And *that*, Robert, is when you're finally going to realize that your life has been a waste. An illusion. Because *love*, Robert...love is what life is all about! Love and sharing." Maureen wrapped her arms around Rob and drew him close against her breasts. Her heart pumped nervously, achingly.

The speakers crackled. It was the voice of the band's female vocalist. "We have a special request tonight to play '*Desperado*' by the *Eagles*. This song goes out to Rob from Maureen. We hope you all enjoy our version of this sentimental rock ballad."

As the music began drifting out over the bar's speaker system, Maureen led Rob in a loving slow dance. One-by-one, other couples joined them on the dance floor, swaying gently to the words and the music.

The song told about emotional barriers, about love, about Rob's future if he didn't face up to his feelings. The song made Maureen cry on Rob's shoulder.

When the ballad was over, Maureen released the hold she had on Rob. She said to him, "If you ever need to talk, you can still come see me at Jackie's place. But after tonight...our relationship is finished." Tears billowed down her cheeks as she turned away from Rob. "Take care of yourself," she whispered, walking away, leaving

Figure 9-5

progress and the public's access to information and ideas. But determining what's fair has puzzled writers, editors, lawyers, and judges for years. Indeed, there is a thin line between fair use and copyright infringement."[36]

In other words, an author must be certain that the copyrighted material he or she is using is indeed considered to be fair use. An attorney specializing in literary matters may need to be consulted. What are the standards used for justifying what is fair and what is unfair? Henderson wrote in *How To Bulletproof Your Manuscript* that, "Conveying the facts and information and advancing the arts and sciences, as well as the dissemination of historical information, would all be considered fair use in most cases. However, when someone's work is used (or usurped) for strictly commercial purposes, the courts have been strict in considering it unfair use, which legally is known as copyright infringement."[37]

What does this mean? Well, using the lyrics in *Black Falcon* would have been strictly for commercial purposes. I would not have been conveying facts or information, nor would I have been advancing the arts or sciences, and I certainly would not have been disseminating historical information. In *How To Get Published - Guaranteed*, however, there were several instances where I either quoted very brief passages (one or two sentences in length) from books and articles, or else I paraphrased the passages into my own words. (The same holds true for *How To Write, Publish & Sell Your Own Aviation Books*, since this book is basically a rewritten version of *How To Get Published - Guaranteed*). In each case, these exemplifications were not used strictly for commercial purposes, but rather they were used to convey important information, facts and figures about writing and publishing, and also to advance the arts. Proper credit was given in all cases to the various authors, thus making them all examples of fair use - except, possibly, the two

Live to Write...Write to Live!

Live to Fly...Fly to Live!

Live to Sail...Sail to Live!

These public service announcements are brought to you by Chevy Alden and Tri-Pacer Press.

Take a moment to pause for a mental health break!

TRI-PACER PRESS, P.O. BOX 840111, Pembroke Pines, Florida 33084 • 305-432-9683
Publisher of Quality Action-Adventure Novels

June 22, 1993

Writer's Digest Books
Permissions Department
1507 Dana Avenue
Cincinnati, Ohio 45207

Dear Permissions Manager:

I am preparing a book, How To Get Published — Guaranteed, to be published in Fall 1993 by Tri-Pacer Press. The publisher estimates first printing as 1,000 copies and the retail price as $16.95.

I'd like nonexclusive world rights to reproduce, in my book under the chapter heading Legal Considerations, the following excerpt from How To Bulletproof Your Manuscript by Bruce Henderson, Copyright 1986, published by Writer's Digest Books:

> "Writers do have the protected privilege of using another's work — with proper credit given — in a 'reasonable manner' without having to secure the permission of the author. This is called 'fair use', and 'fair' is the key word here. Courts have held that such use promotes intellectual progress and the public's access to information and ideas. But determining what's fair has puzzled writers, editors, lawyers, and judges for years. Indeed, there is a thin line between fair use and copyright infringement."
>
> "Conveying the facts and information and advancing the arts and sciences, as well as the dissemination of historical information, would all be considered fair use in most cases. However, when someone's work is used (or usurped) for strictly commercial purposes, the courts have been strict in considering it unfair use, which legally is known as copyright infringement."

If this is satisfactory, please sign and date this letter, specifying any wording you wish to appear as the acknowledgment, and return the letter in the enclosed envelope.

If you do not control these rights, I would appreciate your directing me to the copyright holder. If you have any questions you'd like answered before granting permission, please let me know.

Sincerely,

Richard Cook
aka Chevy Alden

Name Laura A. Smith Date 6-30-93

Wording of Permission from HOW TO BULLETPROOF YOUR MANUSCRIPT copyright © 1986 by Bruce Henderson. Used with permission of Writer's Digest Books, a division of F&W Publications, Inc.

Figure 9-6

lengthy paragraphs I quoted from *How To Bulletproof Your Manuscript*.

Why? Because the amount of material quoted was possibly too long to be considered fair use. Rather than risk a possible court battle to find out if it was fair use, I chose instead to seek permission to use the two passages from Bruce Henderson's book. This was easily done by writing to Writer's Digest Books, the publisher of *How To Bulletproof Your Manuscript*, explaining how I was going to use the two paragraphs. Figure 9-6 shows the permission request I sent to Writer's Digest Books. They signed the permission grant at the bottom of my letter and sent it back to me, which gave me the legal right to use both paragraphs in their entirety.

One point of importance that should be noted here is that song lyrics and poetry are *never* covered under fair use laws, so don't be tempted to use either of these in your writing without first getting permission grants from the copyright holders. I do know of a few self-publishers who have been able to obtain permission grants for song lyrics, so it's definitely possible if you try hard enough (or have the money to hire an attorney to do it for you).

To sum up fair use, an author should protect the rights of other authors, just as he or she would want other authors to protect his or her rights. Play it safe. Play it fair. Play it legal.

CHAPTER TEN

FINANCING

Unless an author is independently wealthy, finding the money to finance a self-publishing venture will probably be the most difficult part of getting a book into print. When I originally set out to self-publish *Black Falcon*, my bank account was not overflowing with great sums of money. What I *did* have, though, was an extra abundance of enthusiasm for my manuscript, lots of ingenuity and many years of experience servicing corporate jets. Together, these three factors helped me finance the publication of *Black Falcon*.

You might be asking yourself, "What does servicing corporate jets have to do with starting a small publishing company?" Well, nothing really - except that corporate jets are the airplanes owned and operated by top business executives, world political leaders, real estate tycoons and other upper-class clientele. I learned from these people that *most* business ventures get financed using other people's money. In other words, you have to borrow money to make money.

Due to my own financial situation (or lack thereof), I knew I would have to borrow money if I was going to get *Black Falcon* into print. After some initial cost estimates for printing and binding 1,000 copies of my novel, I figured I could publish it in a trade paperback edition with a two-color cover for just around $3,000 (plus or minus a few hundred dollars). However, I wanted to go all the way with *Black Falcon* by publishing it in a hardcover edition with a full-color dustjacket. This added another $2,000 to my project, bringing the cost estimate up to $5,000 for 1,000 hardcover copies. Additionally, I didn't own a personal computer at that time, so I had to figure in another $3,000 to cover the cost of professional typesetting. This brought my grand total up to $8,000 for the entire project.

After figuring out the total cost of publishing *Black Falcon*, I calculated my earning potential from the sale of 1,000 books. At a price of $16.95 per book (the original retail price I had set for first printing copies), I had the potential of earning $16,950 ($16.95 x 1,000) before deducting business expenses. The trick was to sell as many books as possible at full retail and to keep my business expenses as low as possible by doing as much of the work as I could on my own. (Note: The actual retail price for the first printing ended up being $17.95 per copy, so I was able to earn a bit more than originally planned.)

With all the figures in my head, I knew it was time to start seeking out a loan. I had $2,500 in the bank that I could use for the publishing venture, so I only needed to borrow $5,500 to start Tri-Pacer Press. The first step in borrowing this money was to write a simple financial proposal explaining the nature of my proposed business, a description of my book, the cost of producing the book, the retail price, the discount price, how I was planning to market the book, and my expected sales revenue from the first printing. Basically, I needed to show potential lenders that I was serious about borrowing the money, as well as serious about paying back the loan.

Next, I showed my financial proposal to family and friends, managing to secure a $5,700 loan this way (payable one year after the publication of *Black Falcon*). I had offers for more money if I needed it, but I never did. With the loan secured, I was ready to launch my company.

Was I worried about not being able to pay back the loan? No, because I had great faith in my book. I also knew that if I didn't self-publish *Black Falcon*, it would never get published - *and I wanted it published more than anything in order to prove that I was a good writer!*

Needless to say, *Black Falcon* got off the ground with great success. I was able to pay back the loan on time, plus cover all of my business expenses, plus cover the cost of the second printing of *Black Falcon*. In other words, Tri-Pacer Press had become self-sufficient right from the beginning.

But what if I hadn't been able to secure the loan? What would I have done then? Well, I still would have self-published *Black Falcon* - just not in a hardcover edition and not with a full-color cover. I also would not have hired a professional typesetter to do the typesetting, but rather I would have sought out a friend with a personal computer and set up the same type of deal I had made with my editor (meaning that payment for services rendered would have been made upon the publication of the second printing of the book). By publishing *Black Falcon* using this method, my production costs would have been in the $3,000 range. I already had $2,500 in the bank, so I only would have needed an additional $500, which I could have earned quickly by taking a part-time job for a couple of months (even if it meant pumping gas into cars or bagging food at the grocery store).

But what if my bank account had been totally empty? What would I have done then? Well, in that case I would have postponed the publication of *Black Falcon* for a year, worked a part-time job, and saved all the money from that part-time job for the eventual publication of my book. Part-time jobs are a great way to earn extra income if you need it. I know this might sound cliched, but where there's a will, there's definitely a way.

Keep in mind that people who *never* give up *will* eventually succeed. I'm a firm believer in positive thinking. This includes *never* taking *no* as the final answer. A good example of this is the sailboat I purchased after doing research for *Black Falcon*. I really couldn't afford to own a 22-foot sailboat at that time, but I went ahead and took out a loan for $2,000 from a bank, searched for an old boat that needed work, bought it, read a couple of books about boat restoration, scrounged for scrap wood in the dumpster behind a yacht interior shop, shopped for marine hardware at a store that specialized in used marine equipment, and turned *Dreamer* into a fine sailing vessel. Of course, I had to eat tuna fish and bologna sandwiches for a long time to pay back the loan, but at least I had my own sailboat. I still own that sailboat today.

Another good example of positive thinking is the sport flying I do out of North Perry Airport. Sport flying is one of the most expensive hobbies a person can get involved with (and one of the most thrilling, if not *the* most thrilling). For a long time after I had earned my private pilot certificate, I couldn't afford to fly. I would get current, fly for a few months until the money ran dry, then quit for a year or two until I saved up some more money, then I'd start the whole process over again. Finally, I just gave it up for good (or so I thought). Then, when *Black Falcon* was published at the end 1988, I thought it was odd that I was an aviation novelist who could no longer afford to fly. Well, where there's a will, there's a way. I became a part-time substitute teacher at a local high school and used all of the money I earned from the teaching job to start flying again. I've been an active sport pilot ever since that time, with not a single on-again off-again cycle. I was financially able to quit my part-time substituting job in 1996 and still afford to fly - the power of positive thinking!

So you see, anything is possible if a person wants it bad enough (whether it be flying, sailing, self-publishing or anything else a person might want to do). *Anything is possible!* Let my experiences be an example to you. Start right now by figuring out what *you* really want out of life, then go figure out how you can obtain it - and *never* take *no* as the final answer.

I'd like to leave you with one more note of inspiration. The cost of the first printing of my second book, *How To Get Published - Guaranteed,* was roughly $4,600 for 1,500 copies, plus $2,000 for typesetting. I learned a lot from the mistakes I had made with *Black Falcon,* so I was really able to cut down my production costs on my second book, while still keeping the quality high. From my $6,600 investment, I received $28,425 worth of books at $18.95 per copy (the retail price for the first printing). An even better cost structure was realized with *How To Write, Publish & Sell Your Own Aviation Books* because I did my own typesetting, so the potential for me to eventually buy my own airplane is definitely out there, folks. Hey, this self-publishing gig is a great thing!

CHAPTER ELEVEN

TYPESETTING

Typesetting is the process of transforming your edited manuscript into camera-ready copy. In other words, pages that are ready to be printed. The steps involved in this transformation include designing page layouts for various sections of the book, selecting type styles and sizes, choosing a typesetter, proofing galleys, paste-up of camera-ready copy and proofing camera-ready copy.

Typesetting is one of the most important stages of book production. Unfortunately, it is also one of the most frustrating parts of the production process (at least it was for me during the typesetting of *Black Falcon*). Any mistakes that make it into your camera-ready copy (such as typographical errors, wrong words, crooked lines of type, etc.) will also make it into your finished book. Because of the number of difficulties that can crop up during typesetting, I will try to guide you through this process by showing you what happened with *Black Falcon*, and then how I avoided these same problems during the typesetting of *How To Get Published - Guaranteed*, and finally how I made things even easier during the typesetting of *How To Write, Publish & Sell Your Own Aviation Books*. (Aren't you glad that I'm giving you the opportunity to learn from my mistakes so that you'll have an easier time of it?) The advice given in this chapter will save you thousands of dollars, because that's how much I saved during the typesetting of this book compared to the cost of typesetting *Black Falcon* and *How To*

Get Published - Guaranteed.

Designing Page Layouts For Your Book. A book can be designed in hundreds of different ways, but the easiest and most cost-effective method is to design your book using one of the half-dozen or so standard industry dimensions (such as 51/2" x 81/2" or 6" x 9" or 81/2" x 11") and to set up the pages using simple, easy-to-read layouts.

Black Falcon was designed using the 51/2" x 81/2" dimensions - a common size for hardcover novels. *How To Get Published - Guaranteed* was originally going to be designed around the 81/2" x 11" format, but the cost estimates I received were way too high. Why? Because most of the cost of producing a book comes from the cost of the raw paper, and a larger book uses a much greater amount of raw paper than does a smaller book. For this reason, *How To Get Published - Guaranteed*, as well as *How To Write, Publish & Sell Your Own Aviation Books*, were both designed using the 51/2" x 81/2" format.

After choosing your book's overall dimensions, you will need to design page layouts for each of the different types of pages that will appear in your book. Every book has many types of pages, and these pages are organized into three distinct sections called front matter, text and back matter. The front matter includes the testimonial page, title page, copyright page, table of contents, list of illustrations, acknowledgments page, dedication page and preface (some of which are not mandatory for every book). The text is the main body of the book. The back matter includes the footnotes, bibliography, index, glossary, appendixes, and any other items a publisher might wish to include.[38]

Designing page layouts is a simple matter of drawing page templates on a piece of paper for each type of page that will be in your book. The easiest way to decide on a design is to go to the library for several hours, look at dozens of books, find one you like and measure all the dimensions. That's what I did. Figure 11-1 shows the page template for the text of *Black Falcon*. Notice that the template indicates the positioning of the page numbers, page headings, chapter headings, first line of text, last line of text, margin placement and any other special instructions I had for the typesetter.

Individual page templates were designed for the title page, copyright page, acknowledgments page and dedication page of *Black Falcon*. Figures 11-2, 11-3, 11-4 and 11-5 show the actual camera-ready copy that was produced by the typesetter from each of these templates. Let's examine each of these pages in detail.

Figure 11-2 shows the title page. Notice that everything is centered and that this page includes the title, author's name, publisher's name, publisher's location, plus an extra bonus I included indicating that the first printing of *Black Falcon* was a Special "First Printing" Collector's Edition. Upon publication of the first printing (which was 1,000 copies) I personally autographed and numbered each copy, thereby making every copy a special investment for the future. Perhaps someday the people who bought those copies will have a valuable first edition sitting on their bookshelf. That is my hope, anyway! For the second printing of *Black Falcon*, as well as for all subsequent printings, I removed the special statement and relocated the author's name to the center of the page. Now turn to the actual title page of *How To Write, Publish & Sell Your Own Aviation Books*. Notice that I listed Walter J. Boyne as being the author of the foreword. This was the first time I had to add additional information to the title page of a book - and well worth it.

Figure 11-3 shows the copyright page from *Black Falcon*, which is probably one of the most important pages of any book. Notice that this page has a justified (aligned) left margin, but an unjustified (nonaligned) right margin. The items usually appearing on the copyright page include dis-

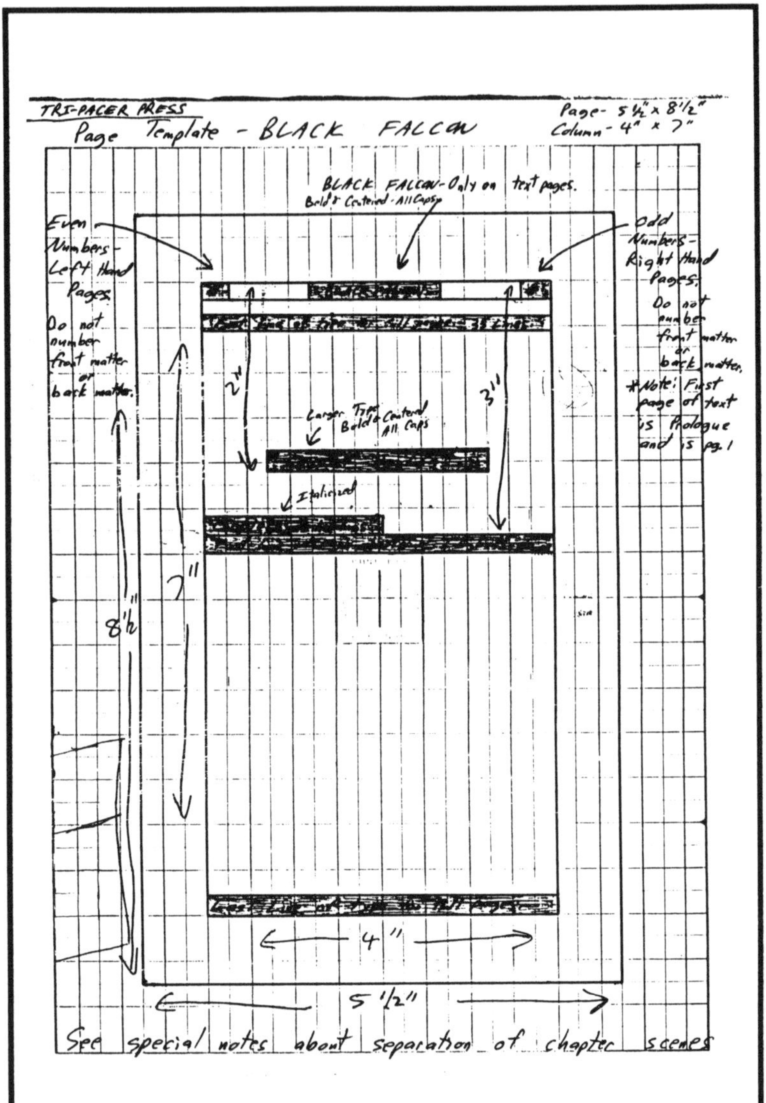
TRI-PACER PRESS
Page Template - BLACK FALCON
Page- 5½" x 8½"
Column- 4" x 7"
BLACK FALCON- Only on text pages.
Bold & Centered - All Caps
Even Numbers- Left Hand Pages.
Do not number front matter or back matter.
Odd Numbers- Right Hand Pages.
Do not number front matter or back matter.
Note: First page of text is Prologue and is pg. 1
2"
3"
Larger Type Bold & Centered All Caps
Italicized
7"
8½"
4"
5½"
See special notes about separation of chapter scenes

Figure 11-1

BLACK FALCON

CHEVY ALDEN

Special "First Printing" Collector's Edition

TRI-PACER PRESS
PEMBROKE PINES, FLORIDA

Figure 11-2

All characters and events depicted in this novel are fictitious creations of the author. Any resemblance to real people, living or dead, or to real events is purely coincidental.

Published by:
TRI-PACER PRESS
P.O. Box 840111
Pembroke Pines, FL 33084

Copyright© 1986, 1988 by Richard Cook
Library of Congress Catalog Card Number: 88-50158
ISBN 0-9619991-5-2

All Rights Reserved. This book may not be reproduced in whole or in part by mimeograph or any other means without written permission from the publisher.

First Printing: October 1988

Printed in the United States of America

Figure 11-3

ACKNOWLEDGMENTS

Special thanks go to *George Diamond, Ph.D.,* whose creative criticism and encouragement kept *BLACK FALCON* a viable project for seven long years.

Special thanks also go to my editor/copyeditor, *Beth Hagan,* whose exceptional faith in *BLACK FALCON* allowed her to take on such a difficult task.

Appreciated thanks go to: My wife, *Bonnie,* who endured many lonely nights while I wrote *BLACK FALCON.* My parents, *Morris* and *Sarina,* who put up with a lot of craziness from me over the years. *Jennifer P. Tucker,* who did a great job on *BLACK FALCON's* cover art. *Humberto (Tony) Miranda,* who provided me with the Spanish translations and slang for *BLACK FALCON. Molly Siobhan McGill,* who lent me her AX-12 electronic typewriter for the final rewrite of *BLACK FALCON.*

Figure 11-4

In memory of George Pagar, who taught me the meaning of freedom.

Figure 11-5

claimer messages, publisher's name and address, copyright statement, Library Of Congress Catalog Card Number, ISBN Number, permission grants, printing date and the country in which the book was printed.

Figure 11-4 shows the acknowledgments page from *Black Falcon*. Notice how I spelled the word acknowledgments. It can also be spelled acknowledgements, with an *e* between the *g* and the *m* - either way is acceptable. This page was set up with justified left and right margins. The acknowledgments page is not necessary, but it is a great way for authors to thank the people who have helped them along the way.

Figure 11-5 shows the dedication page from *Black Falcon*. This page is usually a very personal one for most authors (sometimes with hidden meaning). In *Black Falcon*, I wrote: *In memory of George Pagar, who taught me the meaning of freedom*. George Pagar was the captain of the 76-foot schooner *Freedom*, the guy who taught me how to sail while I was researching *Black Falcon*. George was the one who taught me the meaning of freedom on the open sea and about the glory of wooden sailboats.

You have already seen the page template I made for the text of *Black Falcon* in Figure 11-1. Well, Figure 11-6 shows the actual camera-ready copy that was produced from this page template. Now turn back to Figure 11-1 and compare it to the actual text shown in Figure 11-6. Notice how they match up perfectly. Included at the top of each page of text is the title of the book. This is called a running head and it is usually centered.[39] Also on each page is the page number. This is called a folio and it can either be on the top of the page (like in *Black Falcon*) or at the bottom of the page (like in *How To Write, Publish & Sell Your Own Aviation Books*).[40] One important rule to remember when numbering your pages is that the right-hand pages will be odd-numbered pages and the left hand pages will be even-

BLACK FALCON 1

PROLOGUE

Early morning in mid-February

The gravel-faced captain maintained a firm grip on the Lodestar's control wheel as he hand flew the twin-engined aircraft just fifty feet above the Florida Straits. He was thinking about the rip-off, wondering if it would go down as planned, wondering if he'd have enough guts to pull the trigger of his .357 Magnum. He had never killed a man before. Would it be the same as killing a deer? He wondered a bit more, then put his mind back on flying the airplane.

After scanning the panel, the captain turned his gaze out the windshield. A full moon hung low above the horizon as it made its nightly descent against the western sky. Below, the waves' crests sparkled in the moon's effervescence, creating an eerie, yet tranquilizing feeling for the pilot.

The Lodestar's copilot, meanwhile, was paying no particular attention to nature's artistry, although he, too, was looking out the aircraft's windshield. His eyes were busy darting around, searching in the distant early morning darkness for a certain man-made landmark. He was sure his navigational calculations were correct. Or were they? Maybe he had figured the winds wrong.

Figure 11-6

numbered pages.

While still looking at Figure 11-6, keep in mind that although the camera-ready copy for *Black Falcon* was straight text, some books are packed with many drawings, photos, figures, charts and other odds-and-ends mixed in with the text. *How To Write, Publish & Sell Your Own Aviation Books* is a good example of this. If your book includes lots of extras, simply have your typesetter make borders on the pages that need it, then affix (called paste-up in the printing industry) all the extra items onto the camera-ready copy. Do not, however, affix your photos onto the camera-ready copy. These will be sent to your printer separately.

To paste-up properly, you will need an artist's light table, a transparent gridboard, a transparent T-Square ruler, a transparent 90-degree angle ruler, a roll of non-stick masking tape, an electrically-operated hot-wax roller, a bright table lamp, and a sharp razor or knife. All of these items can be purchased at a good art supply store. These items are shown in the photo in Figure 11-7, spread out on my kitch-

Figure 11-7

en table. The only really expensive item is the artist's light table. They can cost anywhere from $100 up to $500, depending on the quality. I built mine for free and I'll tell you how to build one for yourself. (Don't you just love how I keep giving you all of these money-saving tips?)

To build an artist's light table, you will need an ordinary hood light from an aquarium. Don't buy one - just borrow one from somebody for a few hours. I borrowed the hood light from my daughter's ten-gallon fish tank. (Hey, don't laugh. This method really works. The book you are now reading went through the entire paste-up process using my homebuilt artist's light table.) The hood light was 181/4" in length and 21/2" in height, so I knew that the inside dimensions of my light table had to be at least 181/4" long, 21/2" high, and wide enough to fit the 11" width of my gridboard. The base of the light table was built out of four strips of scrap plywood for the sides and four pieces of scrap hardwood for the corner gussets. The plywood was cut into two 20" x 3" strips and two 12" x 3" strips. I cut the hardwood, which was a long piece of 1" square stock, into four 21/2" long pieces. I placed the four strips of plywood into a rectangular shape, then glued the four hardwood gussets into the four inside corners. After the glue dried, I drilled two pilot holes through the ends of each piece of plywood, extending the holes into the gussets. I then hammered boat nails through the pilot holes, creating a sturdy light table in the process. Then I cut a small 1/2" triangular slot into one side of the light table so that the electrical cord from the hood lamp would easily pass through the side of the light table. And that is how I built my light table.

When I was ready to do the paste-up, I placed my light table on the kitchen table, placed the aquarium hood light inside the wooden framework and passed the cord through the slot in the side of the light table. I then got out an old 11" x 16" picture frame and set it on top of the hood light.

The glass portion of the picture frame made a perfect surface for paste-up work (flat and smooth). The last step was to place the transparent gridboard on top of the picture frame. That's it! My free artist's light table was finished and ready for the paste-up process.

Figure 11-8 shows a photo of the completed light table during the paste-up process, giving you a good idea how the pieces fit together to make a real working light table. To do the actual paste-up work, tape a page of camera-ready copy to the gridboard using non-stick masking tape. (Actually, the tape *does* stick, but it will peel right off the page without tearing the paper when that page is finished.) Make certain when you tape each page of camera-ready copy to the gridboard that the edges of each page are perfectly aligned with the lines printed on the gridboard, otherwise everything you paste-up will be out of alignment in your book and that defeats the whole purpose of using the gridboard. Use the razor or knife to cut out the item that will be affixed to that particular page. Use the hot-wax roller on the back side of

Figure 11-8

the item, being sure to fully cover it with melted wax. Then place the item onto the camera-ready copy, using the two special purpose rulers to aligned the item with the lines on the gridboard. The reason the gridlines appear through the camera-ready copy and the waxed item is because of the light from the fish tank hood, which is why this set-up is called an artist's light board. When you are positive that everything is properly aligned, put a blank sheet of paper over the item to protect it from scratches and then smooth it down firmly using one of the rulers so that the wax flows evenly and adheres the item to the camera-ready copy. In other words, use the ruler on the item as if it was a squeegee removing water from a window. That's it for that page of camera-ready copy. Peel off the tape, place the page someplace safe and repeat the entire process with the next item to be affixed.

Going back to Figure 11-6, notice that the chapter heading is centered on the page. Also notice that certain items are *italicized*, CAPITALIZED and printed in **bold** letters (like in this sentence, for example). These are all ways that typesetters can enhance the look of your text, but it's up to you to tell the typesetter what you want. Most typesetters do not charge extra for these enhancements.

Drawing the page templates is the *only* physical activity that a self-publisher *has* to do during the typesetting phase of book production - unless, of course, he or she has a personal computer and decides to do his or her own typesetting (which is what I did with *How To Write, Publish & Sell Your Own Aviation Books*). If you go ahead and decide to self-publish your book, you already have a jump start on making your page templates because all you have to do is follow my examples. For instance, if you are self-publishing a novel, use the page template from *Black Falcon* as your guide. On the other hand, if you are self-publishing a non-fiction book, use the various page layouts from *How To*

Write, Publish & Sell Your Own Aviation Books as a guide for your page templates. Of course, you are free to design your own page layouts if you wish - just remember that the more complicated you make it, the more expensive it will be. Try to keep your costs as low as possible.

Selecting Type Styles And Sizes. Type styles come in hundreds of different varieties, each with its own name (such as Times Roman, Baskerville, Bookman and Garamond Book). Not all type styles, however, are conducive to easy reading. The four type styles mentioned above *are* easy-to-read varieties that will work nicely for most books. There are others, though, that will work just as well, so ask your typesetter for samples. Most typesetting businesses have charts that show the different type styles they use. Times Roman was the type style used for *Black Falcon*. The type style used for *How To Get Published -Guaranteed* was Times (a minor variation of Times Roman). Both of these typesetting jobs were done by professional typesetters. Times New Roman (another variation of Times Roman) was the type style used for *How To Write, Publish & Sell Your Own Aviation Books*. All three of these type styles are used extensively within the publishing industry and each one of them has variations of itself. For example, the running head shown in Figure 11-6 is Times Roman Bold in a smaller type size, the chapter heading is Times Roman Bold in a larger type size, the chapter sub-heading is Times Roman Italics in the same type size as the text, and the text is just plain Times Roman.

Type size is very important when it comes to a book's readability. The print shouldn't be too small, nor should it be too large. The type size used for the text of *Black Falcon* was 12/14 (pronounced twelve on fourteen in the industry). 12 was the point size measurement (height) of the capital letters plus the spacing above and below the capital letters. 14 indicates that two additional points (a typesetting measurement) were used for extra spacing (leading) between each line of text.[41] Leading adds readability to the text. Common type sizes for books are 9/11, 10/12, 11/13, and 12/14. The type size used for the text of *How To Get Published - Guaranteed* was 11/13, slightly smaller than the 12/14 used for *Black Falcon*. The type size used for *How To Write, Publish & Sell Your Own Aviation Books* was simply 12, because the leading was set automatically by my computer. It might be possible to change the leading on my computer, but I'm still an amateur at this high-tech stuff and haven't yet found the time to seek out an answer.

By choosing a specific type size, you can make a book thinner or thicker (meaning either a lesser or greater number of pages). The smaller the type size, the more words you can fit onto each page - hence, the book will have fewer pages. By going with a larger type size, the less words you can squeeze onto each page - thereby, a thicker book will be created because more pages will be needed to complete the text.

How big (or rather, how small) is a point? To compare the size of a point to a standard measurement, it takes 72 points to equal one inch.[42] As an illustration, the footnote after the previous sentence is 6 points. So you can see that there's not much difference between 11/13 and 12/14, but there is a difference. The reason I chose 11/13 for *How To Get Published - Guaranteed* was because I had so much material to include in the book that if I had used 12/14, the book would have taken too many pages to cover the same material. Those extra pages would have increased the production costs of the book, which in turn would have meant an increase to the retail price. As it is, the book turned out to be 328 pages. If it had been typeset using 12/14, the page count would have been between 350 to 400 pages. I was forced to use 12 for *How To Write, Publish & Sell Your Own Aviation Book* (rather than 11, which is what I wanted to use) because the only type size choices programmed in my computer are 6, 8, 10, 12, 14, 16, 18, 20, 24, 30, 32, 36, 40, and 48. I felt that 10 would have been too small, so I used 12, which brought the final page count of the book up to 376 pages. Figure 11-9 shows a sampling of type sizes.

Choosing A Typesetter. Choosing a typesetter for *Black Falcon* was not difficult, but actually getting the book typeset was an act filled with frustration. In fact, for me it was the most frustrating part of the entire publishing process. To begin the process of finding a typesetter, I decided to go locally so that I could keep close tabs on the progress of the job. I looked up Typesetting in the local Yellow Pages and picked out a dozen companies that advertised custom typesetting services.

Next, I typed a Request For Quotation (RFQ) like the example I had seen in Dan Poynter's book, *The Self-Publishing Manual.*[43] I made a copy for each of the twelve companies, then drove around town delivering the RFQs to the owners of each typesetting shop. Figure 11-10 shows

6 8 10 12 14 16 18 20 24

30 32 36 40 48

Figure 11-9

TRI-PACER PRESS, P.O. BOX 840111, Pembroke Pines, Florida 33084

Publisher of Quality Action-Adventure Novels

REQUEST FOR QUOTATION - TYPESETTING

To: [illegible]

Please quote your best price and delivery for typesetting and paste-up of the following book:

SPECIFICATIONS
Name of book: (Black Falcon)
Type of book: (Fiction)
Total number of pages including front and back matter: (224)
Trim size: (5½ x 8½)
Type style: (12/14 Times Roman)

$7.50/page without paste-up.

QUOTE
Composition, per page: 1680-
Paste-up, per page: 2240-
Delivery charges to our address:
Total price?

$10.00/page with paste-up.

Delivery will be: 21 working days from receipt of manuscript.

Terms: 1/3 order - 1/3 Proofs - 1/3 Delivery

Remarks:

Signed: [signature]

Date: 2/24/88

Any item in this RFQ takes precedence over any industry convention. The manuscript, galley proofs and page proofs shall be returned to customer on completion of job.

Please enclose a copy of your type styles.

Figure 11-10

the RFQ that went to the shop that eventually got the job.

What I got back from the twelve local typesetting shops were quotes ranging from a low of $10 per page to a high of almost $20 per page - a huge price difference! All of the companies estimated the job would take between three to four weeks to complete. Because I was working on an extremely tight budget, I quickly weeded out the higher-priced companies, which narrowed down my choices to the two firms that quoted me $10 per page. As it turned out, I ended up having major problems with both of these companies, which is why I'd rather not mention their names here. Let's just refer to them as Firm 1 and Firm 2.

Firm 1 was my first choice because they had just received a brand new typesetting computer that could produce camera-ready copy without the need for paste-up work on the text. Firm 2 still used an older typesetting computer that required paste-up work (a process that involved cutting up galley proofs and pasting them onto page boards, pasting on running heads and folios, then pasting on any corrections that needed to be made to the text). Remember, this was back in 1988 when newer, high-tech typesetting machines were just becoming available. Many companies still had older equipment. I don't think anyone uses the older machines anymore.

Firm 1 had quoted me a price of $10 per page for clean camera-ready copy, so I stopped by that shop to sign a contract. The owner, however, advised me that his main typesetter had just quit her job and that she was the only person who had been trained on the new typesetting computer, which his firm had just recently purchased. He said he didn't know how long it would be before he'd be hiring somebody who was trained on the new machine. With a scheduled publication date rapidly approaching, I couldn't wait around for Firm 1 to hire someone new, so I walked out the door without signing a typesetting contract.

I immediately drove to Firm 2, where I had also been quoted a price of $10 per page, which broke down to $7.50 per page for the actual typesetting and $2.50 per page to do the paste-up work on the text. I signed a contract with the owner and he introduced me to the woman who would be typesetting *Black Falcon*. The terms of the contract called for the first third of the money to be paid up front as a down-payment, another third upon receipt of the galley proofs, and the final third upon receipt of the finished camera-ready copy. I found this arrangement to be standard among the local typesetting industry in South Florida.

Now take another look at Figure 11-10 and notice that my original page estimate for *Black Falcon* was 224 pages. I had arrived at this page count by first adding up the number of words on an average page of the typed manuscript, then multiplying that number by the total number of typed pages in the manuscript. That gave me a figure of approximately 73,000 words for the entire manuscript. Then I used a chart that had been given to me by one of the typesetting companies which showed different word counts per page for various type styles and sizes. For instance, the chart indicated that a 51/2" x 81/2" page with a 4" x 7" column of text using 12/14 Times Roman would produce 365 words per page. 73,000 words divided by 365 words per page equaled 200 pages. I then counted in the front matter and back matter, plus added in some extra pages to play it safe, and came up with an estimate of 224 pages. What I didn't know at the time was that not all typesetting computers are set up the same way, meaning that one computer might produce 365 words per page for a given set of dimensions, while another computer might be set up to only produce 345 words per page for those same parameters. Anyway, as it turned out, the final page count for *Black Falcon* came to 257 pages of text, plus 7 pages of front and back matter, for a total of 264 pages.

Let's now take a look at how the increased page count affected the overall cost of my typesetting job. The original quote was for $2,240 (which was derived by multiplying 224 pages by $10 per page). This was almost $800 less than the $3,000 cost estimate I had come up with back when I was still trying to figure out how to finance my self-publishing venture, so I figured I was ahead of the game. The increase to 264 pages boosted the final cost of typesetting up to $2,640 (an additional $400), which was still less than what I had originally planned on spending, so I was still ahead of the game.

Now go back to Figure 11-10 once more and take a look at the statement that says: Delivery will be 21 working days from receipt of manuscript. Working days at Firm 2 were Monday thru Friday, so I fully expected the job to be finished in about four weeks. The manuscript was delivered on April 29, 1988, which meant that the job should have been completed by June 1, 1988. That would have given me plenty of time to send the camera-ready copy to my printer and still meet my proposed publication date of August 1st.

Unfortunately, the June 1, 1988 delivery date came and went. In fact, Firm 2 hadn't even touched my manuscript during the first three weeks they had it. I was furious because they had already received a down-payment for the job and I was receiving nothing in return. When I questioned the owner about his company's lack of progress, I was told that my job was a very big undertaking for them and that they were busy meeting deadlines for local condominium newsletters, local advertising flyers, local business forms and other small typesetting jobs that had tighter deadlines than my book. In other words, because my job was so large, it was getting pushed aside to make room for quick and easy jobs.

Well, I explained that I had a deadline, too, and that I would take my business elsewhere if they didn't begin my job immediately. And so, Firm 2 finally began typesetting *Black Falcon*.

Proofing The Galleys. Galley proofs used to be long sheets of typeset text that had not yet been set into individual pages. Nowadays, with the newer typesetting computers, galley proofs are usually already in page format. The purpose of galleys is to allow you to proofread the text for errors so that any necessary corrections can be made prior to sending the camera-ready copy to the printer.

Firm 2 delivered my galley proofs (the long sheets) during the first week of July (five weeks after the typesetting had begun and eight weeks after the manuscript had been delivered). Figure 11-11 shows an actual sheet from the galley proofs that eventually became pages 24 and 25 of *Black Falcon*. Notice all the mistakes the typesetter had made on just this one sheet. Altogether, I found more than 600 errors in the galleys, so I returned them for corrections.

After spending two weeks correcting the mistakes, Firm 2 gave the galleys back to me for additional proofing. This time I found close to 125 mistakes. Once again, the galleys were sent back for corrections.

When I received the galleys for the third time, the error count was down to around 60, the fourth time 25, the fifth time 15, and finally, after a month of proofing and sending back the galleys, I signed them off as being correct and ready for paste-up.

Paste-Up Of Camera-Ready Copy. In order to make camera-ready copy from galley proofs, the text must be divided into individual pages, then pasted onto sheets of thin cardboard using hot-wax. Folios (page numbers) are then added, as are running heads and anything else that might be required for each particular page. This portion of book production is done by a paste-up specialist who usually works for the typesetting company (although this entire process is becoming outdated due to the newer typesetting computers,

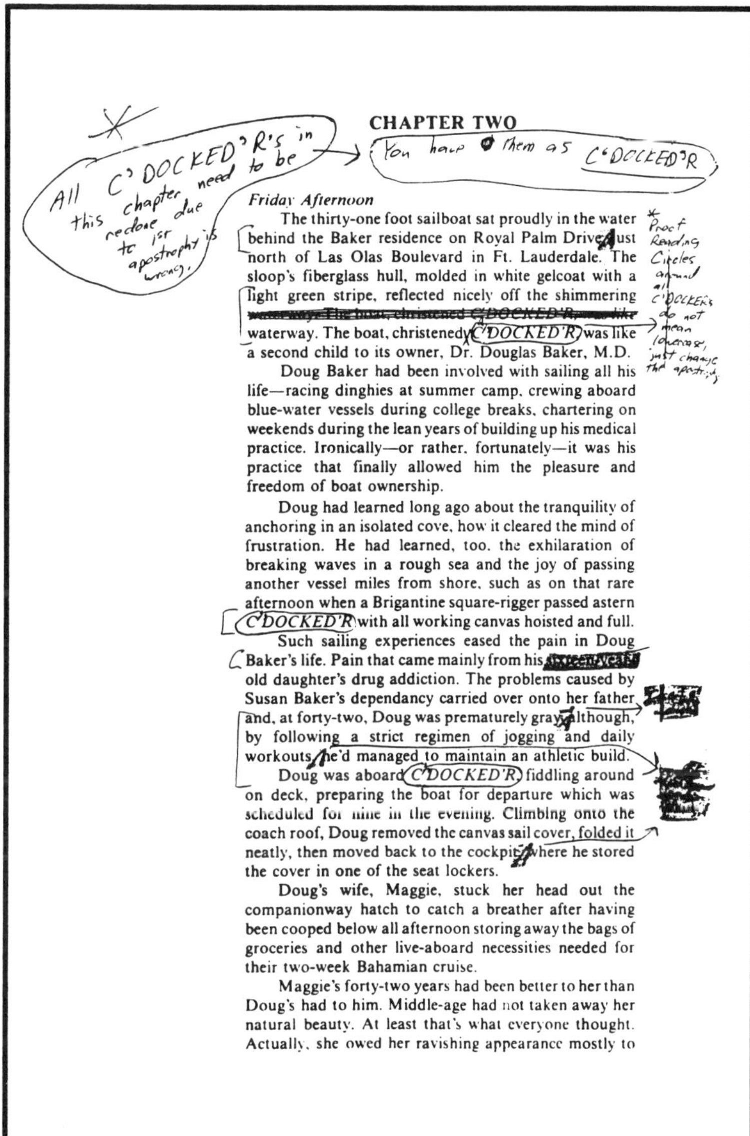

CHAPTER TWO

Friday Afternoon

The thirty-one foot sailboat sat proudly in the water behind the Baker residence on Royal Palm Drive, just north of Las Olas Boulevard in Ft. Lauderdale. The sloop's fiberglass hull, molded in white gelcoat with a light green stripe, reflected nicely off the shimmering waterway. The boat, christened *C'DOCKED'R* was like a second child to its owner, Dr. Douglas Baker, M.D.

Doug Baker had been involved with sailing all his life—racing dinghies at summer camp, crewing aboard blue-water vessels during college breaks, chartering on weekends during the lean years of building up his medical practice. Ironically—or rather, fortunately—it was his practice that finally allowed him the pleasure and freedom of boat ownership.

Doug had learned long ago about the tranquility of anchoring in an isolated cove, how it cleared the mind of frustration. He had learned, too, the exhilaration of breaking waves in a rough sea and the joy of passing another vessel miles from shore, such as on that rare afternoon when a Brigantine square-rigger passed astern *C'DOCKED'R* with all working canvas hoisted and full.

Such sailing experiences eased the pain in Doug Baker's life. Pain that came mainly from his old daughter's drug addiction. The problems caused by Susan Baker's dependancy carried over onto her father and, at forty-two, Doug was prematurely gray, although, by following a strict regimen of jogging and daily workouts, he'd managed to maintain an athletic build.

Doug was aboard *C'DOCKED'R* fiddling around on deck, preparing the boat for departure which was scheduled for nine in the evening. Climbing onto the coach roof, Doug removed the canvas sail cover, folded it neatly, then moved back to the cockpit, where he stored the cover in one of the seat lockers.

Doug's wife, Maggie, stuck her head out the companionway hatch to catch a breather after having been cooped below all afternoon storing away the bags of groceries and other live-aboard necessities needed for their two-week Bahamian cruise.

Maggie's forty-two years had been better to her than Doug's had to him. Middle-age had not taken away her natural beauty. At least that's what everyone thought. Actually, she owed her ravishing appearance mostly to

Figure 11-11

and in fact was not used for either *How To Get Published - Guaranteed* or *How To Write, Publish & Sell Your Own Aviation Books*, except for the diagrams, figures and illustrations). When everything is pasted into position on the boards, photocopies are then made so that final proofing can be done without marking up the actual camera-ready copy.

Proofing The Camera-Ready Copy. When I received the photocopies of the camera-ready copy, I drove down to the Florida Keys by myself, checked into a quiet motel for the weekend and spent the entire time holed up in my room reading over the page proofs. Basically, everything was correct. Of course, by that time I had read the galleys so many times that I completely missed about a dozen typographical errors in the camera-ready copy, all of which ended up in the book. Actually, this wasn't so bad for a first-time publisher, especially when you consider that even major publishing houses usually have at least a dozen typos in their books. The only complaint I had about the camera-ready copy was that all of the corrections that had been made to the text created shadows on the photocopies of the page proofs. Apparently, instead of correcting the mistakes in the computer during the galley stage, the typesetter had just typed entirely new lines of text wherever the mistakes occurred and then pasted these corrected lines of type directly over the lines that contained the mistakes. I had been proofing lightened photocopies during the galley stage, so I didn't know the corrections were being pasted on top of the actual text, causing a layer cake effect, which caused the shadows to appear on the darkened photocopies of the camera-ready copy. You can see these shadows quite clearly in Figure 11-12, which shows a page of camera-ready copy from *Black Falcon*.

I contacted my book printer as soon as I got back to town and told him about the shadows on the photocopies.

48 BLACK FALCON

After taking care of her other customers, Maureen came over. "What was that, Nick?"

"Another Heiney will do fine, darlin'. And a smooch." He puckered his mouth, causing the Marlboro to dangle precariously from his lips. Nick was drunk.

"Sorry, Nicky. Just sold the last one."

"Last one? We talkin' kiss or Heiney here?"

"Kiss."

"Yeah, until Rob gets here. Then you'll be all over him like a little puppy dog."

She ignored the rude statement, turning away instead to get Nick his beer. When she returned with the chilled bottle of Heineken, she caught a glimpse of Rob walking in through the tavern's side entrance. She waved to Rob, but apparently he didn't see her.

Maureen watched as he strolled over towards the juke box, deposited two quarters and punched up several songs. She guessed they were probably Southern Rock and Roll tunes. Rob liked that kind of music—*The Marshall Tucker Band, Lynyrd Skynyrd, Molly Hatchet, The Allman Brothers.* Jackie always kept a few of their songs in the juke as a special favor to Rob.

Maureen kept her eyes upon him as he walked up to the bar counter. "Hey, cutie," she said cheerfully, greeting him as he sat down on the stool next to where Nick was standing.

"Hi." Rob leaned across the counter and gave her a soft kiss on the cheek. He looked into Maureen's eyes, then at her smile. "How're things going tonight? Looks pretty slow for a Saturday."

"It is. Jackie says some of the guys from EuroAir are working overtime to fix a maintenance problem on one of their Treestars. I think that's what he called it."

"Tri-Star," Rob corrected. "A Lockheed L-1011."

"Yeah, that was it! You're so smart, Rob. You know

Figure 11-12

(Basically, I told him my typesetter had possibly screwed up the job.) I asked him if the shadows would appear on the printed pages of *Black Falcon*. He immediately set my mind at ease by saying that the shadows could be obliterated by performing an additional filming procedure called double-opaquing. In fact, he said it was no problem at all and that it was a common occurrence with pasted-up copy. The extra cost for doing the additional procedure amounted to only $80, so it was money well spent when the time came to actually print the books. Once my printer assured me that the typesetter's sloppy work was not a problem, I paid the typesetter the balance of the bill, picked up the actual sheets of camera-ready copy, and shipped them off to the printer. Things were finally coming together!

Solving The Typesetting Problem. By the time *How To Get Published - Guaranteed* was ready to be typeset, astronomical advancements had been made in the field of personal computers. Excellent typesetting programs, in conjunction with high-quality inkjet and laser printers, brought desktop typesetting within easy reach of anyone who knew how to operate a computer. With such equipment, a person could save thousands of dollars by producing their own professional-looking camera-ready copy. Unfortunately, I was not one of those people. You see, I was still computer illiterate, having written and edited both *Black Falcon* and *How To Get Published - Guaranteed* by hand. In fact, I was petrified of computer technology.

Anyway, because of the difficulty that had occurred during the typesetting of *Black Falcon* using a local firm, I decided to take a different approach for the typesetting of *How To Get Published - Guaranteed*. My decision was to join the desktop typesetting bandwagon by finding someone who owned a personal computer and asking that person to do the typesetting for me. Diane Tysenn, who worked with me at the airport, dabbled in computers as a hobby. I asked

Diane if she would typeset my book, offering her $2,000 for the job, plus an additional $500 bonus if she'd wait for the entire payment until after the book was in print. Although she had never typeset a book before, she agreed to give it a try. Figure 11-13 shows the contract I signed with her.

I'll have to say this about Diane - she gave it a good try for someone who had never used a typesetting program before! Unfortunately, Diane was so involved with outside interests that she had little time left over in which to learn the intricacies of her typesetting software - although she did achieve some success. Figure 11-14 shows the formatting she was trying to accomplish using the page layout I had designed. We mutually agreed that it would be best if we voided our contract on friendly terms, and for me to find somebody else to do the typesetting. I'm sure that if we both had had more time to wait out the learning curve, Diane would have done a fine job typesetting *How To Get Published - Guaranteed.* I mean, Figure 11-14 proves she was on the right track - and I knew lots of self-publishers who had successfully typeset their own books, so I knew that personal computers were the wave of the future. It just wasn't the right answer for me at that time.

So how did I finally solve my typesetting dilemma? Well, I called BookMasters in Ashland, Ohio, and inquired about their professional typesetting service. The rough quote I received over the telephone for *How To Get Published - Guaranteed* was anywhere from $6 to $8 per page for a 4" x 7" column of text, depending on the type size and complexity of the page layout. I immediately faxed BookMasters the first couple of pages from my manuscript, along with a specification sheet and an RFQ. When I received the written quote, it was for $6 per page, which came to $1,632 for an estimated page count of 272 pages - a lot less expensive than the typesetting job that was done on *Black Falcon.* Figure 11-15 shows the written quote that was sent to me

TRI-PACER PRESS, P.O. BOX 840111, Pembroke Pines, Florida 33084 • 305-432-9683
Publisher of Quality Action-Adventure Novels

CONTRACT FOR TYPESETTING

This agreement between Tri-Pacer Press (also known as Richard Cook) and Diane Tysenn is entered into for the purpose of typesetting the following manuscript:

How To Get Published — Guaranteed.

It is understood by both parties that this manuscript was written by Chevy Alden (also known as Richard Cook) and that the copyright holder of this work shall be the author. As an unpublished work, the copyright date shall be 1992 & 1993, with a proposed publication by Tri-Pacer Press in 1993.

Tri-Pacer Press agrees to pay Diane Tysenn the sum of $2,000 for typesetting this manuscript, plus a bonus of $500 for accepting full payment after this book is published and has sold enough copies to cover the typesetting fees. This shall amount to sales of approximately 300 - 350 copies.

Diane Tysenn, as the typesetter, agrees to keep the contents of this manuscript confidential until after the publication of the book.

Diane Tysenn, as the typesetter, agrees to typeset the manuscript in page form and to make any corrections necessary upon proof-reading by Chevy Alden.

This agreement is entered into on 12/10/92 by the following parties:

Richard Cook as Tri-Pacer Press [signature] Date 12/10/92
Diane Tysenn [signature] Date 12/10/92

Voided

Figure 11-13

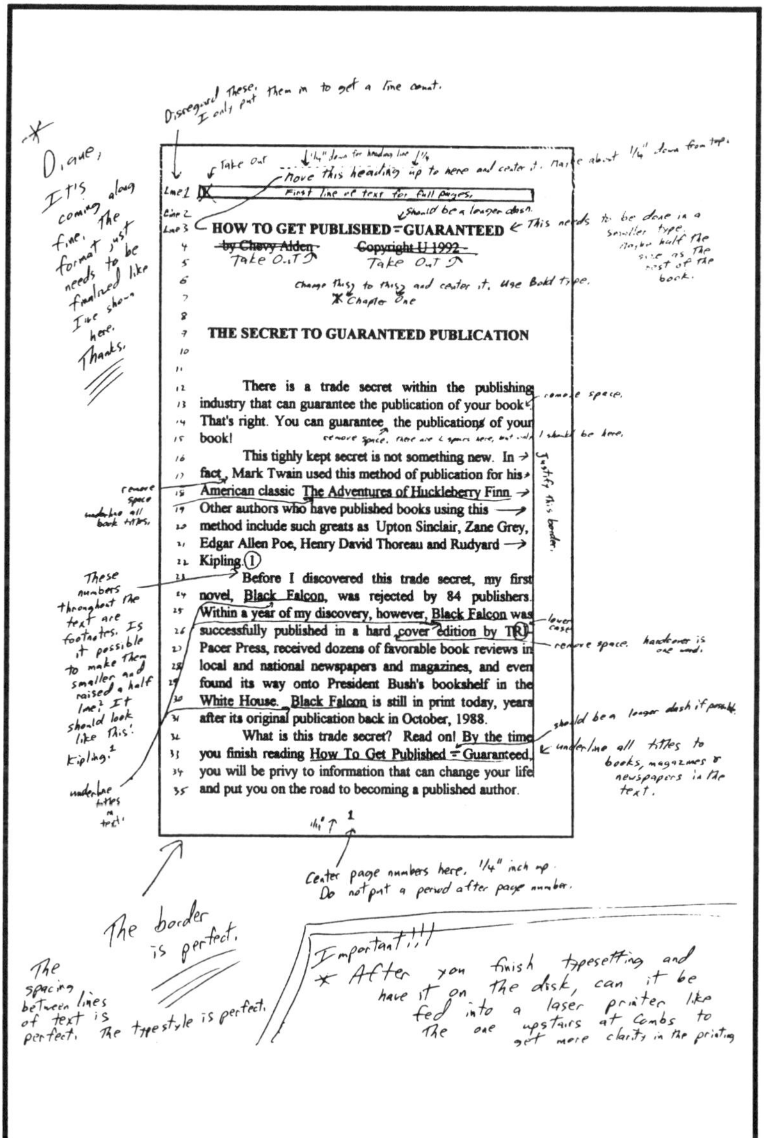

HOW TO GET PUBLISHED - GUARANTEED

~~by Chevy Alden~~ ~~Copyright U 1992~~

THE SECRET TO GUARANTEED PUBLICATION

There is a trade secret within the publishing industry that can guarantee the publication of your book! That's right. You can guarantee the publication of your book!

This tighly kept secret is not something new. In fact, Mark Twain used this method of publication for his American classic The Adventures of Huckleberry Finn. Other authors who have published books using this method include such greats as Upton Sinclair, Zane Grey, Edgar Allen Poe, Henry David Thoreau and Rudyard Kipling.(1)

Before I discovered this trade secret, my first novel, Black Falcon, was rejected by 84 publishers. Within a year of my discovery, however, Black Falcon was successfully published in a hard cover edition by TRI-Pacer Press, received dozens of favorable book reviews in local and national newspapers and magazines, and even found its way onto President Bush's bookshelf in the White House. Black Falcon is still in print today, years after its original publication back in October, 1988.

What is this trade secret? Read on! By the time you finish reading How To Get Published - Guaranteed, you will be privy to information that can change your life and put you on the road to becoming a published author.

1

Figure 11-14

BookMasters, Inc.
638 Jefferson St., P.O. Box 159
Ashland, Ohio 44805
800/537-6727 In Ohio call 419/289-6051 FAX 419/281-1731

July 16, 1993

Richard Cook
Tri-Paper Press
P.O. Box 840111
Pembroke Pines, FL 33084

Dear Richard:

Thank you for the opportunity to quote on your latest book. Following is our quotation for the HOW TO GET PUBLISHED-GUARANTEED.

Book Trim:	5 3/8 X 8 1/2
Text Size:	11/13 X 24
Estimated Page Count:	272
Condition of Manuscript:	Customer to furnish moderately edited manuscript.
Illustrations:	----
Runningheads:	Yes
Folios:	Yes
Endnotes:	----
Tabular Material:	----
Type Faces:	Times
*Proofs:	Rough pages - .15 each Pages - .15 each Revised pages - .15 each
Additional Items:	Author's alterations $1.85 per line affected Postage - AT COST Stats - $5.00
Terms:	Net 10 days
Please Note:	Editing and final proofreading are the responsibility of the customer unless otherwise stated

*BookMasters suggests that you do not prepare your index from rough pages as pagination may change.

An additional charge will be requested if finished pages are required for first pass.

Prices:	Composition through pages $6.00 per page

We look forward to working with you.

Sincerely,

Raymond F. Sevin
VP/General Manager

RFS/lm

Figure 11-15

by BookMasters.

Not only were they less expensive, but they were far more professional as well. In fact, BookMasters concentrated solely on book production (no business forms, condominium newsletters, advertising flyers, etc.) and their equipment included state-of-the-art computers that produced camera-ready copy without the need for paste-up work. When I received sample copies of the first four pages of *How To Get Published - Guaranteed*, the page proofs were perfect, except for the dashes in the text, which were elongated by the computer (and which were easy to fix).

Needless to say, I gave BookMasters the job of typesetting *How To Get Published - Guaranteed*. They did an excellent job - and without any headaches or frustrations! BookMasters had solved my typesetting problems.

But how did I find out about BookMasters in the first place? After all, they were located in Ashland, Ohio (now in Mansfield after relocating from Ashland in 1996). Well, to properly answer that question I'll have to backtrack to 1987 when I first read *The Self-Publishing Manual*. Inside Poynter's book was a list of top-quality book printers. One of them was BookMasters, so I wrote to them seeking prices for the printing and binding of 1,000 hardcover copies of *Black Falcon*. Unfortunately, BookMasters only produced paperback books at the time, so *Black Falcon* was out of the question. BookMasters informed me, however, that they were the national sales representative for another book printer called BookCrafters that *did* produce hardcover books. I was also told that both companies provided professional typesetting services at reasonable prices.

I had already decided to go locally with my typesetting job, figuring that the cost and quality would be the same everywhere in the country (the biggest mistake I ever made during my entire publishing career), so I never sought out a typesetting quote from either company. I *did*, however, ask for a printing and binding quote from BookCrafters, the book printer associated with BookMasters. Well, the way things worked out, BookCrafters ended up getting the job of producing *Black Falcon* (the first printing at their plant in Chelsea, Michigan, and the second and third printings at their plant in Fredericksburg, Virginia).

Even though I didn't use BookMasters for the production of *Black Falcon*, I never forgot the friendly treatment I had received from one of their corporate executives. On many occasions, I had called him up on his toll-free telephone number asking for advice. He was always willing to give it to me (for free). So when Diane and I decided to back down from our typesetting agreement, I gave BookMasters a call to see what they could do for me in the way of typesetting - which turned out to be a lot. BookMasters really did make life easier for me during the typesetting of my second book. However, due to a higher page count (no fault of BookMasters), the final charges came to roughly $2,000.

And now for the true facts about how I learned how to do my own typesetting. While my second book was being printed and bound, my wife insisted that she was not going to sit around quietly and watch me write another book by hand. She insisted that Tri-Pacer Press was going to buy a computer and that I was going to learn how to use it. Well, I was finally forced to give in to technology and I'm glad to say, "Thanks, Bonnie, for forcing me to seek out a much easier and less expensive way!"

I bought a Packard Bell Legend 2033 computer with a Hewlett Packard DeskJet 520 printer. The book you are now reading was totally written and typeset by myself using this equipment. The cost for typesetting *How To Write, Publish & Sell Your Own Aviation Books* came to $0.00 - that's right, absolutely *free*.

I am not a computer expert (far from it, since I still

Writer's Book Mart Newsletter

The Newsletter For Authors Who Want To Get Published

Motto: Writing Is Fun — Special Edition - Page 1

Guaranteed Publication

The Secret Is Revealed

Our featured book review this issue is the award-winning self-help manual "How To Get Published - Guaranteed"

Every writer has heard the infamous *Catch-22* of the publishing world that begins with the proverb, "If you don't have an agent, you can't get published," and ends with, "If you aren't already published, you can't get an agent."

Chevy Alden faced this grim situation after completing his first novel, the air-sea, action-thriller *Black Falcon*. After receiving rejection letters from 84 publishers and 18 agents, Alden decided to research the publishing industry as if he was a detective working on a life or death case.

Within a year *Black Falcon* was successfully published in a hardcover edition and received favorable reviews from dozens of magazines and newspapers. *Waterfront News* commented that Alden wrote "convincingly about cruising the Bahamas aboard a small sailing yacht." *Airliners* magazine called the numerous flying sequences "exciting and realistic." Other trade publications wrote that *Black Falcon* was "a real page turner" and "hard to put down." *The Miami Herald* simply called Alden's smuggling yarn "a down-and-dirty tale." All of this and more for a book that was originally rejected by 84 publishers.

How did Alden finally succeed? His secret is revealed in *How To Get Published - Guaranteed*, an autobiographical account of his meandering through the publishing labyrinth.

Witty and entertaining, this self-help manual delves deeply into the entire publishing process including writing, editing, copyright law, cover design, typesetting, printing, binding, promotion and marketing. Alden also explores the emotional side of writing and publishing, a topic that other how-to books leave out.

Although the text is somewhat technical, Alden presents the material in an easily readable and understandable manner, almost as if he were sitting around a cozy fireplace telling his readers how they could assure the publication of their books.

How To Get Published - Guaranteed is highly informative and enriched with more than 100 photos, documents, illustrations, publishing forms and sample letters. Destined to become a classic, this book is a *must-read* for any author who wants to learn about publishing from an insider's point-of-view.

Chevy Alden Does It All

Author, Pilot, Sailor

Writer's Book Mart Newsletter met up with Chevy Alden at North Perry Airport, his favorite hangout when he's not writing or working on his sailboat.

WBMN: What is your career background?

ALDEN: I've worked as an aircraft refueler since 1976, but because I have expensive hobbies, sport flying and sailing, I've also had to work a second job as a high school substitute teacher.

WBMN: Now that *Black Falcon* is in its third printing and *How To Get Published - Guaranteed* has won an NAIP Book Award, what's next?

ALDEN: Well, I've got about a dozen books planned for the future. That's enough writing to keep me busy for the rest of my life. Two of the books will be sequels to *Black Falcon* which will make my fictional air-sea adventure series a trilogy. The other books will all be nonfiction.

WBMN: What are the highlights of your writing career?

ALDEN: Actually, there are five events that stand out. The first was being chosen as an *Honored Guest*, along with Roxanne Pulitzer and F. Lee Bailey, for the *3rd Annual Palm Beach Literary Forum & Author's Luncheon*. The second was receiving a personal letter from President George Bush informing me that he had a copy of *Black Falcon* in the White House. The third was when *Black Falcon* sold out of its first printing. The fourth was when *How To Get Published - Guaranteed* won an *NAIP Book Award*. And the fifth was when I received a *Certificate of Merit for Outstanding Accomplishment* from *Writer's Digest* magazine.

WBMN: Were you always interested in writing?

ALDEN: No. In fact, when I was in high school I hated grammar and composition. It wasn't until I read Ernest K. Gann's autobiography *A Hostage To Fortune* that I decided to give writing a try. Gann was a pilot, sailor and novelist. He lived the life of an adventurer. I set up a 10-year goal to follow in Gann's footsteps and that's exactly what I've done. Today I'm a pilot, sailor, author and adventurer just like Gann was.

WBMN: When you made your decision to become an author, did you enroll in any writing courses?

ALDEN: No. I developed my own unique style of writing. It's been described as "folksy" and "down-to-earth." Basically, I write like I talk. From all the correspondence I've received in the mail, this is one of the

Writer's Book Mart Newsletter — Special Edition - Page 2

main reasons why people enjoy reading my books.

WBMN: You've written and published both fiction and nonfiction. Do you have a preference?

ALDEN: Definitely fiction, because it allows me to use my imagination and be creative. It involves fantasy and escapism from the real world.

WBMN: What prompted you to write *Black Falcon*?

ALDEN: I had a need to solve some personal problems, so I did it through the characters in my novel. It worked.

WBMN: And what about *How To Get Published - Guaranteed*? Why did you write this book?

ALDEN: The idea was actually given to me by somebody in attendance at one of my speaking engagements. I had just finished talking about the 84 rejection letters I received for *Black Falcon* and how I discovered the secret to guaranteed publication. During a question and answer session someone suggested that a book based on my personal experiences and research would be a big seller. That person was correct. You see, publishing is nothing more than a game with a set of unwritten rules. Unless an author learns these rules, he or she will never get published. According to industry statistics, over 4,500 manuscripts get written by unpublished authors for every one that becomes a book. This means that for every 250 new authors who break into the world of publishing, more than one million will remain outsiders. It doesn't have to be this way. That's why I wrote *How To Get Published - Guaranteed*.

WBMN: To what do you attribute your success as an author?

ALDEN: Lots of perseverance and an unwillingness to accept *no* as an answer. I'm also a firm believer that anyone can achieve their goals no matter how big if they are willing to make sacrifices. For example, the only way I'm able to afford sport flying is by working a second job. Another example is *Dreamer*, my 22-foot sailboat. When I bought her, all I could afford was an old hulk that needed lots of attention. After reading a couple of books about boat restoration, I fixed *Dreamer* into a nice coastal cruiser. A lot of the parts that went into her came from marine salvage yards. That's what I mean by *sacrifices*. I have used this same *can-do* attitude with my writing career.

WBMN: Now that you've achieved your goal of emulating the lifestyle of Ernest K. Gann, do you have any future goals?

ALDEN: To be able to write on a full-time basis. Hopefully soon.

WBMN: Do you have any advice for writers and wannabe writers?

ALDEN: Write from your heart. Write what you know. Write every day, even if it's just an entry in a journal or diary.

Featured Book Review

How To Get Published - Guaranteed

Writer's Book Mart Catalog

The Writer's One-Stop Shop For Writing Resources!

How To Get Published - Guaranteed (by Chevy Alden) An award-winning, self-help manual that describes the publishing process from an insider's point-of-view. ... Winner of the NAIP Book Awards competition. (328 pages, Trade Paperback, $20.95 postpaid)

Is There A Book Inside You? (by Mandy Bingham and Dan Poynter) A self-help manual that describes how to pick your topic, research it and write about it. A fantastic writing reference guide. Chosen as a Writer's Digest Book Club main selection. (236 pages, Trade Paperback, $16.95 postpaid)

1001 Ways To Market Your Books (by John Kremer) This guide features over 1,000 ideas, tips and suggestions for marketing your books. ... A great resource book for publicity-minded authors. (480 pages, Trade Paperback, $21.95 postpaid)

Black Falcon (by Chevy Alden) This action-adventure novel is a flashback to a time when Florida's nighttime sky was a dangerous place to be, the early 1980s ... Written by the author of "How To Get Published - Guaranteed". (237 pages, Hardcover, $20.95 postpaid)

Ordering Information: Check or Money Order only (payable to Tri-Pacer Press). Florida addresses add 6% sales tax. Send to: Writer's Book Mart Catalog, c/o Tri-Pacer Press, Post Office Box 840111-N, Pembroke Pines, FL 33084-2111

Writer's Book Mart Newsletter
A Division Of Tri-Pacer Press
Post Office Box 840111-N
Pembroke Pines, FL 33084-2111

The Newsletter For Authors Who Want To Get Published

Figure 11-16

don't know how to use the Internet or E-mail), but I do know that a personal computer is the way to go if you want to be a self-publisher. Plus, the cost of a computer and its associated equipment is fully tax deductible as a business expense. Not only can you typeset your own books, but you can also typeset your own press releases, brochures and newsletters. Figure 11-16 shows a two-sided newsletter I produced as an advertisement for my books. It was simple, fast and free! As you can tell, I am now sold on the advantages of modern technology.

CHAPTER TWELVE

ARTWORK, PHOTOGRAPHS & COVER DESIGN

Without an eye-catching cover design, it is doubtful that you will sell very many copies of your book, even if your text is of bestseller quality. Think about the last time you purchased a book. Did you bother reading any of the text prior to your purchase? If you did, you're an exception to the rule. Marketing studies have shown that the average person spends less than twenty-five seconds checking out a book.[44] They'll start out by looking at the front cover. If they like what they see, they'll read a portion of the back cover. It's at that point that most people put the book down. If they are still interested, however, they'll read the entire back cover. Then, if they're still interested in the book, they'll read the inside front and back flaps (if it's a hardcover book). At that point, they'll make their decision about whether or not to buy the book.

A book's cover design can make or break a book. Therefore, if you are going to self-publish, don't skimp or short-cut the development of your cover design in any way. What does this mean? Well, if you're wealthy, you can hire a commercial artist to design a cover for your book. That's simple enough. But what if you're like me and don't have $2,000 to $3,000 to plunk down on a professional artist? Don't fear! Within this chapter I will tell you how I hired a commercial artist to design a professional-looking book cover for *Black Falcon* for only $100, but first I'd like to discuss several odds and ends regarding cover design.

The color *red* sells best![45] Why do you think I chose this color for the cover of *How To Write, Publish & Sell Your Own Aviation Books*? Go into any bookstore and see how many publishers use red in their cover designs. This doesn't mean that red should be the *only* color used for book covers, though. For instance, the primary color on the cover of *Black Falcon* is black, which symbolizes death, suspense and danger - all very appropriate for a novel dealing with South Florida's deadly narcotics trade.

The front cover should include the book's title and subtitle (if it has one), plus the author's name. If the book's title doesn't specifically explain what the book is about, and if the book doesn't have a subtitle, then a catchy phrase should be used to help attract people's attention. The cover of *Black Falcon*, for example, includes the catchy phrase *A Thrilling Air-Sea Adventure Novel*, thereby attracting the attention of pilots and sailors. For the cover of *How To Get Published - Guaranteed*, I added the subtitle *A Self-Help Manual For Assuring The Publication Of Your Books*. For my third book, *How To Write, Publish & Sell Your Own Aviation Books*, I used neither a catchy phrase nor a subtitle because the title fully explains what the book is about. I did, however, include the name of Walter J. Boyne as the author of the foreword because his name adds great credibility to the book.

When choosing type sizes for the front cover, don't let your ego get carried away by printing your name in larger letters than your book's title. Sure, it's done for big-name authors like Robert Ludlum and Sidney Sheldon, but they are mega-authors who can sell their books by name recognition alone. Most authors can't do this.

The cover design of self-published books should meet the industry standards of the genre under which they fall. Novels, for instance, usually require full-color (also called four-color) artwork for their covers. To break this down even more specifically, westerns usually require artwork of rugged cowboys riding on horses, whereas aviation novels require artwork of airplanes. The same can be said for all books - fiction and nonfiction. Each genre requires its own specific type of cover design, whether the book is a spy novel, how-to manual, medical journal, sports book, history book, biography, etc., so be sure to match the cover design of your book to its own genre.

The front cover of *Black Falcon* has artwork of a sail-boat and an airplane, plus a map of Florida - all three of which clearly target the book's main audience (people who enjoy reading air-sea action novels about Florida's illegal smuggling trade). For *How To Write, Publish & Sell Your Own Aviation Books*, I chose a more conservative cover design, yet it is lively and up-beat and gives a very positive message. Notice also that the cover design for *How To Write, Publish & Sell Your Own Aviation Books* has a photograph rather than artwork. This is one of the primary differences between fiction and nonfiction. A nonfiction book does not need to have artwork in order to sell the book as long as the overall design fits within industry standards. Rather, a nonfiction cover should be highly informative, especially the cover of a how-to book. I did go all out on the cover photo by renting a photographic studio for a one-hour time block and having the photo shot by a professional photographer after I had arranged all the items I wanted in the picture. Figure 12-1 shows this being done.

The back cover of a book should be used as a billboard - a place to tell potential readers all about your book. Plastering a large photograph of yourself on the back cover is a waste of valuable free advertising space. Use this space to your advantage. If you want to use a photograph of yourself, make it small enough so that it won't overpower your printed message. Better yet, place the photo on the inside of the back cover. That's what I did with *Black Falcon*. With

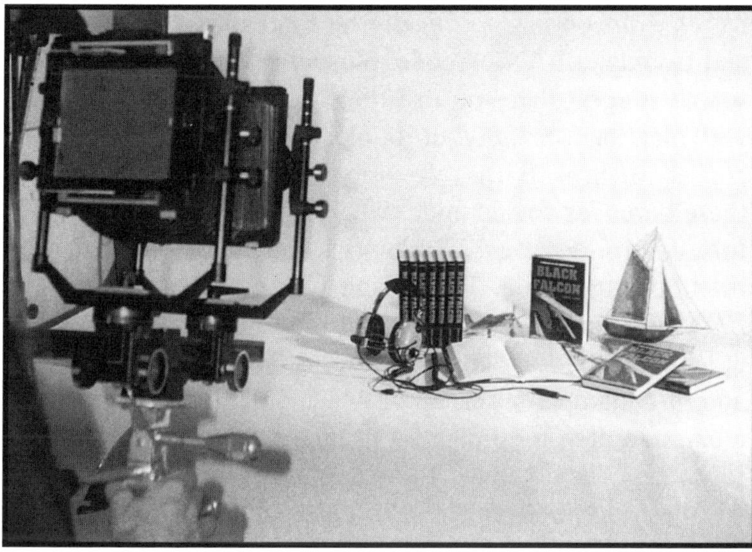

Figure 12-1

How To Write, Publish & Sell Your Own Aviation Books, my photo appears within the actual text of this chapter. To tell you the truth, most people don't care what the author of a book looks like. Photographs are mostly used to appease an author's ego.

The spine of your cover should not be overlooked, especially since most books are displayed spine out, which means that the front cover won't even be visible to customers unless they physically remove the book from the shelf. The spine should have the title of the book and your name.

Several options can be added to the cover of your book, depending on how and where you intend on selling it. These options include retail price, ISBN Number, Bookland EAN barcode, blurbs (the industry term for testimonials) and a short biography about yourself (usually no more than a single paragraph). For *Black Falcon*, I chose not to include the retail price or Bookland EAN barcode, but I did include all of the other options. For *How To Write, Publish*

& Sell Your Own Aviation Books, I included the retail price and Bookland EAN barcode, plus all of the other options. I would suggest that you use the Bookland EAN barcode if you plan on selling your books to the major bookstore chains, such as Barnes & Noble, because the chains use these barcodes for computerized buying, selling and inventory control. Figure 12-2 shows a sample of a barcode for the bookstore trade. The reason I didn't use a barcode for *Black Falcon* is because most of the sales for this book are at aviation and sailing supply houses, air shows, boat shows and through mail-order, all of which don't require the use of a barcode. If you do plan on using a barcode, they are easy to obtain. The company I use is Fotel (ex-GGX) located at 41 West Home Avenue, Villa Park, Illinois, 60181. If you send them the title of your book, the ISBN Number and the retail price, they'll send you the barcode on film within 48 hours, which is then sent to your printer later on in the publishing process. It's as simple as that and the cost is less then $25. If you use Fotel to obtain your barcodes, ask them to

ISBN 0-9619991-6-0
EAN
9 780961 999162
51895>

Figure 12-2

send you a copy of their booklet *Machine-Readable Coding Guidelines For The U.S. Book Industry*. This free booklet will explain in detail everything you need to know about barcodes.

And now it's time for me to tell you how I managed to get the cover of *Black Falcon* designed for a mere $100. All it really took was a little ingenuity. When I first started out on the cover design for my novel, I knew I couldn't afford to hire a commercial artist, so I came up with two rough ideas using simple line drawings - a process that even an inartistic person like myself could handle easily, although this type of cover design definitely would not meet the industry standards for an action-adventure novel. The two cover designs I had sketched out are shown in Figure 12-3 (actually, they were traced out). Both designs used a map of Florida with blood spilling out from Lake Okeechobee, signifying the hundreds of murders associated with South Florida's illegal narcotics trade. On one of the covers I used a DC-4 airfreighter, signifying another illegal drug flight into

Figure 12-3

South Florida's airspace. The other cover had a MAC-10 automatic pistol, the weapon of choice among drug smugglers during the early 1980s.

The more I studied my two designs, the more I realized I was not an artist. Open for ideas, I suddenly came upon a solution - the Art Institute of Fort Lauderdale. This local college-level art school was ripe with commercial art students who were hungry for practical experience and exposure. On the other hand, I was hungry for somebody who was trained in commercial art and who was willing to work for a minimum of pay. It was a natural pairing.

I set up a meeting with the administrators at the art school and explained to them about Tri-Pacer Press and how I was going to self-publish my novel. I explained the concept of self-publishing and that it all revolved around working with a small budget. They gave me the go-ahead to place a Freelance Job Order on the school's job-posting bulletin board. Figures 12-4 and 12-5 show the Freelance Job Order that I posted. On this form, I briefly described the project and pay rate. I also included a section for students to sign up so that I could interview them and see samples of their artwork.

On the day I held the interviews (January 22, 1988), four students showed up with their portfolios. I explained to each one that although I knew nothing about commercial art, I knew whether or not I liked a particular piece of artwork. I also explained that the job would require the chosen artist to paint an aircraft in full detail. One of the art students, Jennifer Tucker, told me that her father flew seaplanes for a local airline that flew between Florida and the Bahamas and that she had painted a picture of a corporate jet taking off at night. When she showed me the air-brushed painting, I knew right then and there that Jennifer had won the job. Not only did she have the ability to paint an airplane, she had a personal interest in aviation as well.

Figure 12-6 shows the Freelance Contract that I signed with Jennifer Tucker. Her job was to provide several rough sketches (called thumbnails), a rough watercolor painting of the chosen sketch, an airbrushed painting of the final cover art, and a layout board of the entire cover (called a cover mechanical). In return, she would be compensated with a check for $100, her name in the book as the cover designer and one complimentary copy of *Black Falcon*. But there was more! Jennifer decided to use this freelance job as her senior class project (for which she eventually received an *A*). Then, after graduation, she used her copy of *Black Falcon* as part of her portfolio while looking for a job as a commercial artist. How many art students can leave college with the experience of having successfully designed the cover of a published book? Not many! Jennifer eventually landed a job in the Graphics Department at Zerox. The last time I saw her (which was at the Miami Book Fair International in 1991), she had gained enough experience to open her own design studio in South Florida.

Figures 12-7 and 12-8 show the eight thumbnails that Jennifer had designed for me. Before Jennifer had started work on the thumbnails, I laid out three requirements for the cover art - an old, prop-driven airfreighter had to be depicted, the airplane had to be in a flight profile, and either Florida or smuggling had to be a major theme in the picture. After choosing thumbnail #8, shown in Figure 12-8, I had Jennifer add a sailboat to the drawing so that the cover would appeal to sailors as well as to pilots (since a lot of the story revolved around sailing in the Bahamas). This addition produced thumbnail #9, the final cover concept, which can be seen in Figure 12-9.

The next step for Jennifer was to produce a watercolor painting of the cover artwork. She did this painting on a large piece of cardboard, which provided plenty of margin space on which to write instructions for any changes or

FREELANCE JOB ORDER

DATE: 12/15/87 MAJOR: Commercial Art

Circle one: (Student) Graduate Both

COMPANY: Tri-Pacer Press/aka-Richard Cook/aka-Chevy Alden

ADDRESS: P.O. Box 840111

Pembroke Pines, Florida ZIP: 33084

PHONE: No phone calls - letters only CONTACT: Richard Cook

JOB DESCRIPTION: Attention: Commercial Art Students!

* ← This book jacket would not sell many novels! Design one for me. I know what I want, but my art is with words — not pictures. Tri-Pacer Press is a very small publishing company set up by myself → over

*See Reverse for more details.

BUDGET: $100 + Students name printed on inside front flap of every book jacket under heading: Jacket Designed by + One autographed copy of BLACK FALCON DEADLINE: March 15, 1988

DATE	STUDENT	MAJOR	CONTACT DATE	INTERVIEWED BY	DECLINED JOB
1/22	[illegible]	[illegible]	[illegible]	[illegible]	[illegible]
1/22	[illegible]	[illegible]	[illegible]	[illegible]	[illegible]
1/22	[illegible]	[illegible]	[illegible]	[illegible]	[illegible]
1/22	[illegible]	[illegible]	[illegible]	[illegible]	[illegible]

****

NAME OF STUDENT/GRAD DOING JOB: Jennifer Tucker

MONEY RECEIVED: $100.00

COMMENTS:

Figure 12-4

to publish BLACK FALCON, an adventure novel that takes place in South Florida. Writing under the pseudonym Chevy Alden, BLACK FALCON has taken me seven years to complete. I now need an artist to help with the jacket design— something that will jump out and say, "BUY ME! READ ME!" This project will not only be exciting, it may also lead to national exposure for your art. — Your name will be printed on every book jacket. The jacket design will also be used for all of my advertising and promotional material. If you are interested, please read the "back jacket copy" which I have provided. This will give you a general idea as to BLACK FALCON's theme. I have also provided an excerpt from Chapter One.

I will be holding interviews for this project on the following dates and times: 11:30-1:00 January 22nd, in room To Be Assigned.

______ January ______, in room ______.

______ January ______, in room ______.

Please sign up with Donne Evans and mail me a short note stating your name and preferred interview date. I will not be contacting you, so please show up on the day you are scheduled. Specific times are not necessary. Interviews will be on a first come - first taken basis. A final decision will not be made until after all signees have had their chance. *Bring your portfolio and dress casually. GOOD LUCK! Chevy Alden - "Live Your Dreams!"

BLACK FALCON is a violent flashback to a time when South Florida's nightime sky was a shadowy place to be...the early 1980's...the boom years for aerial marijuana smuggling. Drugs were cool...sex was loose...life was easy. Instant wealth was offered to any pilot willing to risk his life flying tramp airfreighters down to Colombia and back.

One of those outlaw airmen was Rob Jensen, a physically strong but emotionally unstable doper pilot who decided to quit the drug trade at the wrong time. With nobody to turn to for help, he entered into a desperate struggle for survival against his distraught emotions, nature's elements and the deadly world of drug running.

Follow along as this fast-paced adventure novel takes you from a sleazy Miami bar to a famous Key West saloon, from the cockpit of a drug plane to a clandestine airfield in the mountains of Colombia, from a tropical Bahamian cruise aboard a private sailboat to a climactic action-packed ending on Ft. Lauderdale's Venice-like canal system.

Figure 12-5

FREE LANCE CONTRACT
To Be Completed Jointly By Artist And Client

Date 1/22/88
Name of Client(please print) Tri-Pacer Press, c/o Richard Cook
Client's telephone# [redacted]
Name of artist(please print) Jennifer Tucker
Artist's telephone# [redacted]

Description of Job (include # of pieces of finished art or prints)
Artist to produce commercial quality design, layout and final artwork for jacket cover for the novel BLACK FALCON. Several type styles and sizes will be tested for the jacket's lettering. Client to make final approval on art and lettering decisions. (1) copy of finished art required.

CHECKLIST (check where applicable)

ARTWORK OR PRINTS:

1. Number of pieces of finished prints/key art (1)
   A. Does Client keep all thumbnails? YES X NO ___
   B. Does Client keep all contact proofs sheets/layouts? YES X NO ___
   C. Does artist receive printed work for portfolio? YES X NO ___ (1) copy of book & jacket cover with name printed on inside front flap of jacket cover.

Dates of Completion (if applicable)
A. Thumbnails 2/12, 1988
B. Layouts/proofs 3/11, 1988
C. Finished art 3/25, 1988

AGREEMENT OF CHARGES AND PAYMENT:

1. **COST:**
   Artwork $100.00
   Supplies N/A (if not included in cost of artwork)
2. **PAYMENT:**
   Check ___ Cash X
   Supply costs to be advanced? YES ___ NO X
   Final payment due upon receipt of finished work by client? YES X NO ___ if no, other arrangements ___
3. **PRINTING RIGHTS:** (if applicable)
   ~~Does client get one time printing rights? YES XXXX NO XXXX~~
   Does client get complete printing rights? YES X NO ___

Artwork will be property of Tri-Pacer Press for copyright. Artwork will be used for jacket cover and promotional use.

SIGNATURE CLIENT [signature] **DATE** 1/22/88

SIGNATURE ARTIST Jennifer Tucker **DATE** 1/22/88

All thumbnails, layouts, proofs and finished artwork will be returned to Tri-Pacer Press, c/o Richard Cook.

Figure 12-6

corrections that were needed. Figure 12-10 shows the watercolor painting after the changes and corrections had been indicated by me. Then it was time for Jennifer to paint the airbrushed version of the cover artwork.

While Jennifer was doing the airbrushing, I was fully occupied with other aspects of the cover. Figure 12-11 shows the photograph that was chosen for the back flap of the dustjacket. This was one of about twenty black and white photographs that my wife had taken of me at North Perry Airport. I was also busy writing the back cover copy, as well as the copy for the author biography, which would appear on the back flap of the dustjacket. Instead of writing new copy for the front flap of the dustjacket, I chose to use the first eight paragraphs from Chapter One. I did this because I felt that these paragraphs were strong enough to hook readers into buying the novel. I then went to my typesetter and had them typeset the cover copy in various type styles and sizes. After making photocopies of the cover art, the author photo and the typeset cover copy, I made a prepublication mock-up of the dustjacket by pasting all of the above items onto a piece of cardboard. Figure 12-12 shows the dustjacket mock-up.

When Jennifer finished the airbrushed cover art, she presented it to me for final approval. Figure 12-13 shows the results of Jennifer's work. Notice that the airbrushed painting does not have any lettering on it. Instead, the top of the painting was sprayed solid black. You see, during the printing process, artwork remains separate from the cover copy (lettering) until the final stages of production, at which point the cover copy and the artwork are meshed together to produce the actual book cover. It's all very technical and you don't really have to know the mechanics of it, just that you should keep the artwork separate from actual cover copy when it is all sent to the printer.

The next step in designing a book cover is to produce a

Figure 12-7

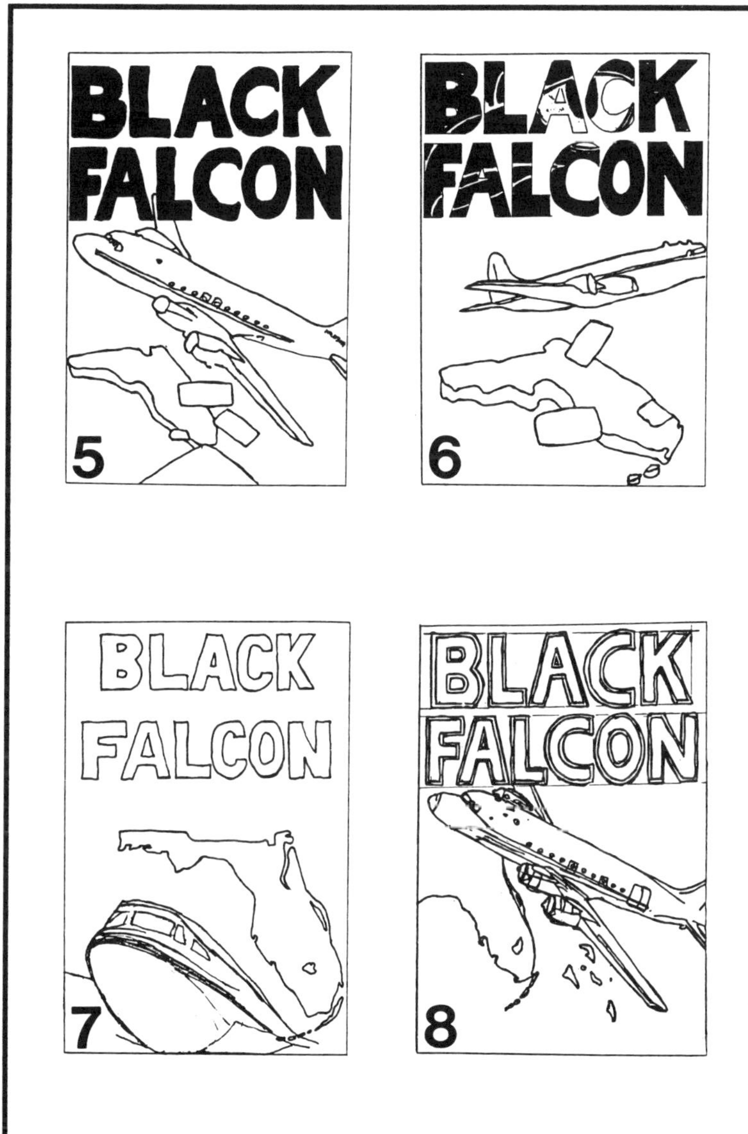

Figure 12-8

Figure 12-9

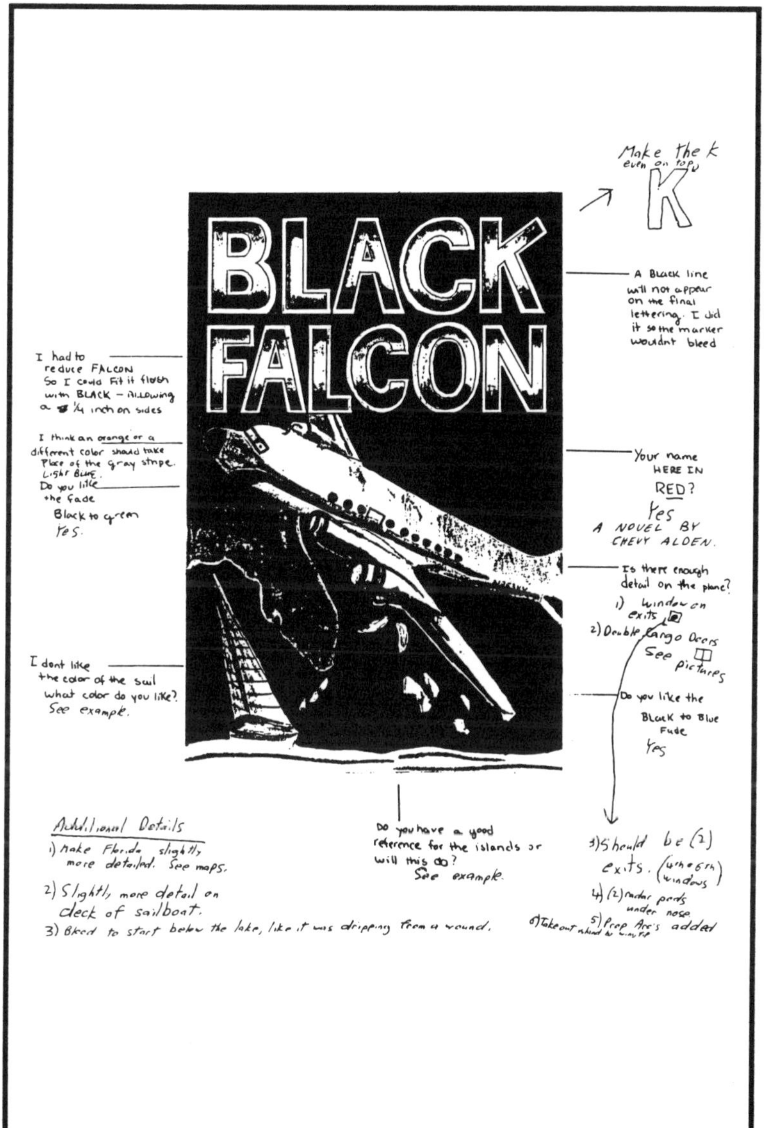
Make the K even on top
K
BLACK FALCON
A Black line will not appear on the final lettering. I did it so the marker wouldn't bleed
I had to reduce FALCON So I could Fit it flush with BLACK - allowing a 1/4 inch on sides
I think an orange or a different color should take Place of the gray stripe. Light Blue.
Do you like the fade Black to green yes.
Your name HERE IN RED? Yes
A NOVEL BY CHEVY ALDEN.
Is there enough detail on the plane?
1) window on exits
2) Double Cargo Doors See pictures
I dont like the color of the sail what color do you like? See example.
Do you like the Black to Blue Fade Yes
Additional Details
1) Make Florida slightly more detailed. See maps.
2) Slightly more detail on deck of sailboat.
3) Bleed to start below the lake, like it was dripping from a wound.
Do you have a good reference for the islands or will this do? See example.
3) Should be (2) exits. (4th & 6th windows)
4) (2) radar pods under nose.
5) Prop Arc's added

Figure 12-10

Figure 12-11

Figure 12-12

Figure 12-13

cover mechanical, which is sort of like a blueprint of the cover copy and artwork. In order to do this, all of the dimensions of the finished book must be known. These dimensions can be obtained easily by talking to your printer after your page count is finalized. One item that will make this procedure even easier is a Dustjacket Layout Sheet (or a Cover Layout Sheet if the book will not be using a dustjacket). Most printers can provide you with one of these forms, usually at no cost. (In fact, printers can provide you with a lot of free help just for the asking. Self-publishing is a growing business and they are eager to help newcomers in order to gain their business.) Figure 12-14 shows the Dustjacket Layout Sheet that I used for *Black Falcon*. It should be noted that the dustjacket of a hardcover book has many more dimensions than the cover of a paperback book. Your printer can help you figure out the exact dimensions you will need for your book cover. These dimensions *must* be correct, otherwise the cover will be misaligned. This, in turn, will ruin the total appearance of your book. Take the

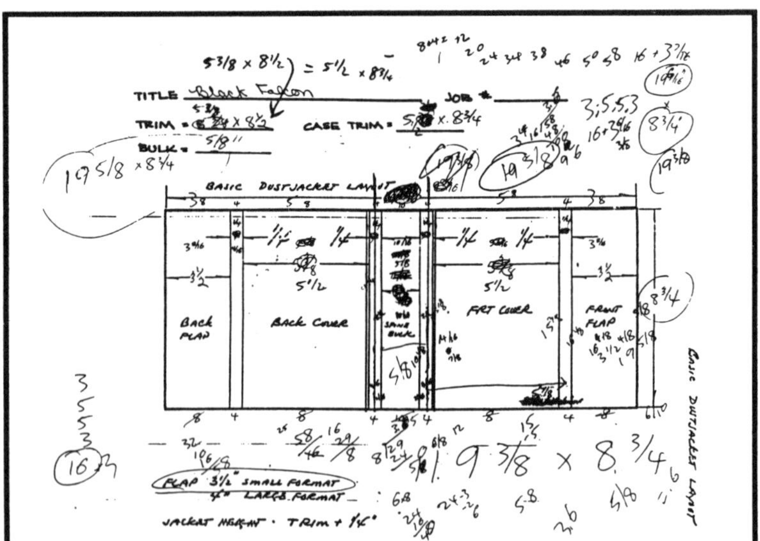

Figure 12-14

time at this stage of the game to double check all of your measurements. In fact, triple check your measurements.

Here is how I figured out the dimensions for the cover of *Black Falcon*. First of all, I had a choice of making the front and back flaps of the dustjacket either 3 1/2" wide or 4" wide. I chose the smaller measurement. Next, a 1/4" spacer had to be added to each flap to compensate for the extra paper that would be used when the dustjacket got wrapped around the front and back covers of the hardcover book. Next, I had to add in the dimensions of the front and back covers, both of which were 5 1/2" wide. Next, I had to add in two separate 1/4" spacers to compensate for the extra paper that would be used when the dustjacket got wrapped around from the front and back covers to the spine. Finally, I had to add in 5/8" to compensate for the spine width (called spine bulk). Spine bulk depends on the thickness of the paper being used for the text. This thickness is described as PPI (pages per inch) and the correct figure can be obtained from your printer. In order to get my final dimension for the width of the entire dustjacket, I added up the various measurements and arrived at a figure of 19 5/8".

For the height of the dustjacket, I had to add an extra 1/4" to the book's 8 1/2" height measurement due to a 1/8" overhang at the top of the hardcover binding and another 1/8" overhang at the bottom of the hardcover binding. These measurements were given to me by my printer, which gave the dustjacket an overall height of 8 3/4". (I'll admit that this mathematical figuring uses a lot of fraction conversions, so if you're not good with fractions, get some help from somebody who is. *It is imperative that all of your cover dimensions be correct!*)

After all of my figuring, the final dimensions came out to 19 5/8" x 8 3/4", plus an additional 1/4" border all around the cover mechanical (called a bleed). Figure 12-15 shows the pasted-up cover mechanical. The 1/4" bleed is clearly

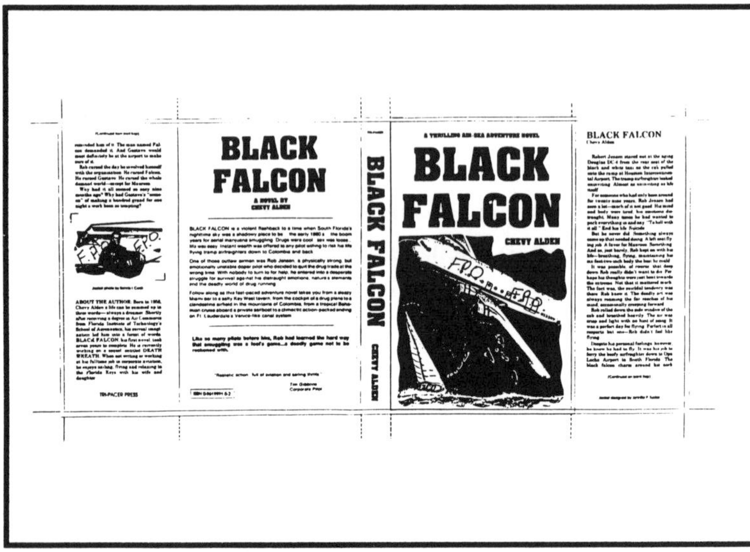

Figure 12-15

visible as the lines around the edges of the cover. Be sure to draw all lines on the cover mechanical using an artist's non-reproducing blue pen or pencil (also called non-photo blue). Any good art supply store carries these items.

Another point of interest in Figure 12-15 is the fact that the original cover artwork and the original author photograph were not pasted onto the cover mechanical. Rather, photocopies of these two items were pasted on merely as references for the printer. Notice, too, the changes in the type styles and sizes that were used for the final cover copy shown in Figure 12-15 as compared to the type styles and sizes that were used on the pre-publication mock-up shown in Figure 12-12. I added spice to the cover's appearance by using a type style known as Megaron Bold for the title of the book and the author's name.

Now take a look at Figure 12-16, which shows the mechanical overlay. Because this overlay is so critical to the outcome of a book cover, I have mounted Figure 12-16 sideways so that you can see it in a larger magnification. I

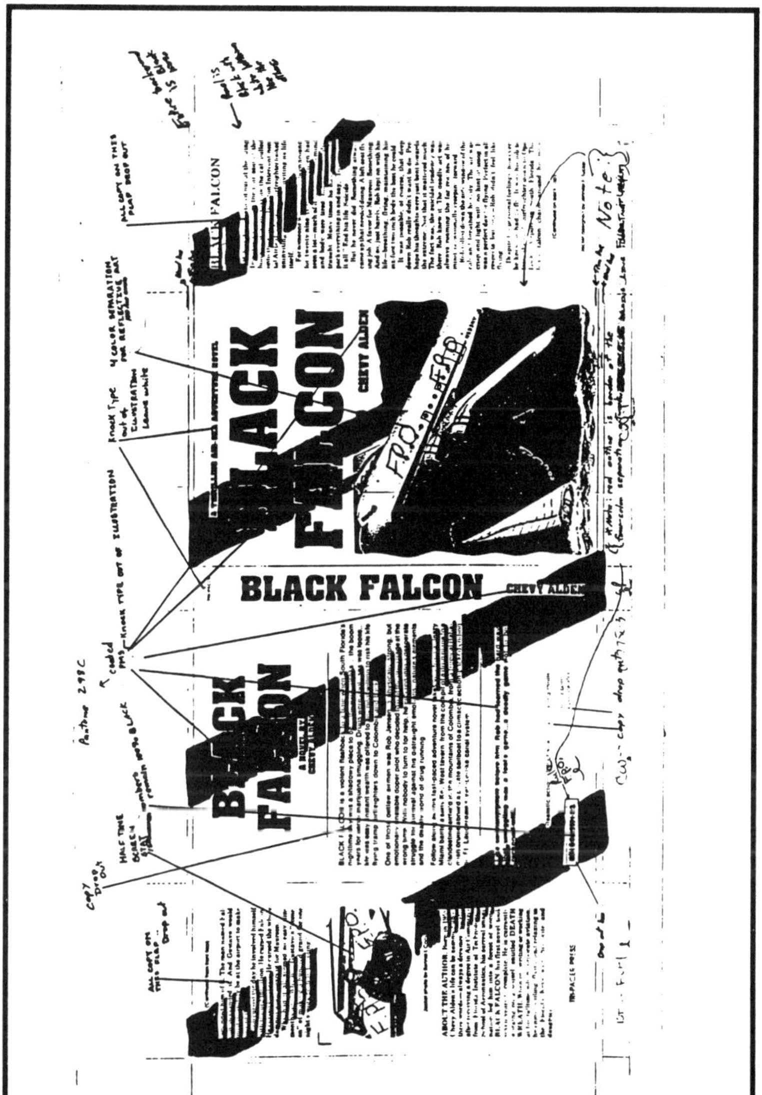

Figure 12-16

did not know that I needed a mechanical overlay, nor did I even know what one was, so I was fortunate to have hired an artist who knew her job well and who provided me with the mechanical overlay as part of the project even though it was not mentioned in the contract. The mechanical overlay is a clear sheet of tracing paper that gets taped over the cover mechanical for the purpose of passing along specific instructions to the printer.

For instance, the copy on the cover mechanical is black lettering on a white background. For the actual dustjacket of *Black Falcon*, however, I wanted these colors to be reversed. In other words, I wanted white lettering on a black background, which brings up another point. There are more shades of black in the printing business other than just black. You can have 50% black, 70% black, and so on. I wanted 100% black for the background of the dustjacket. This was indicated on the mechanical overlay with the words *Entire background is black 100%*. And now back to the original problem of reversing the black and white during the printing process. Jennifer knew that what I wanted was for the printer to print the background using black ink, but not to print the lettering at all so that it would appear white on the dustjacket (the color of the raw paper stock). She indicated this to the printer on the mechanical overlay by coloring in black diagonal lines with white spaces around the lettering. She also wrote the words *All copy...drop out*. The black diagonal lines told the printer to use black ink for the background. The words *drop out* told the printer not to print the particular items indicated - in this case, the cover copy. In other words, the cover copy was not printed at all. What appears to be white printed lettering on the dustjacket is actually the unprinted white paper stock beneath the printed black background - sort of like a ghost image of the cover copy. This all might seem confusing to you, but commercial art students are trained in this kind of stuff.

The instructions were different for the printing of the ISBN Number on the lower left corner of the back cover. In that case, I needed the background to remain white and the ISBN Number to be black. Jennifer drew a rectangular box around the ISBN Number and wrote the words *Drop out box* and *Numbers remain 100% black*. These instructions told the printer exactly what to do. The end result was that the ISBN Number appeared as a series of black numbers within a white box surrounded by the black background of the dustjacket.

A variation of the drop out process is *knock out*. This term can be seen used in Figure 12-16 in relation to the book's title and the author's name. In each of these cases, the lettering would be dropped out of the black background, but would then be printed over using a different color (as opposed to leaving it the color of the white paper stock).

To indicate the exact color I wanted for the knocked out items, I went to the Art Institute of Fort Lauderdale and bought a PMS color chip. Most well-stocked art supply stores carry these PMS chips. PMS stands for the Pantone Matching System, which is an ink coloration system used in the printing industry to indicate approximately 500 different color shades that can be mixed using different combinations of inks. Each shade has its own number. Also, there are different numbers for coated and uncoated paper stocks. I wanted a particular shade of blue for the knocked out items, so I looked through the Pantone chips in the school's art supply store until I found the exact color I wanted. The number for this particular shade was Pantone 298C. On the mechanical overlay, Jennifer wrote the instructions *Coated PMS* and *Pantone 298C*, indicating to the printer the exact shade of blue I wanted for the title of *Black Falcon*.

Next, look at the instructions on the mechanical overlay in Figure 12-16 for the black and white author's photograph that was pasted onto the cover mechanical. The instructions were *Halftone Screen 150*. A halftone is a photograph that has been filmed through a special screen that transforms the image into thousands of tiny dots of various sizes.[46] The more dots per inch, the clearer the printed photo will be in the book. Because I wanted my picture to be very clear, I chose 150 dots per inch (called 150-line screen). Halftones, however, can be produced with less clarity using a 65-, 85-, 120-, or 133-line screen.[47] Just remember that the larger the number, the sharper the image. By the way, this is the same process you would use for any black and white photographs appearing in the text of your book. The cost is roughly $10 per halftone. Color photographs are a different story, with each one having to go through the four-color separation process, which can add thousands of dollars to your final printing bill. Fortunately, black and white halftones can be made from color photographs, however, they are not as clear as halftones made from black and white photos.[48] All of the halftones throughout *How To Write, Publish & Sell Your Own Aviation Books* were made using 150-line screens from color photographs.

The last item you should take notice of on the overlay in Figure 12-16 is the artwork for the front cover. I purposely left the artwork for last because it's the most important marketing tool a publisher has for selling a book. Think of the front cover as a billboard, a marquee, a platform, a stage on which to display a book. A lousy cover will mean lousy sales, even if the book is a masterpiece. Do you remember what I said at the beginning of this chapter about a book cover having to meeting industry standards? You *must* meet these standards if you intend to sell your book. I intended on selling *Black Falcon* when I first decided to self-publish my novel and Jennifer's artwork has helped me to do just that on a continuing basis. Her cover design has definitely met industry standards for action-adventure novels.

To convey to my printer what I wanted done with the

airbrushed painting, Jennifer wrote instructions on the mechanical overlay that said *4-Color Separation For Reflective Art*. This told the printer that the artwork would need to go through a computerized process known as four-color separation that would break down the painting into four separate images on film - red, blue, yellow and black. This, in turn, meant that the production of the dustjackets would need four separate press runs.[49]

Although the four-color process is very costly, it is a necessity for most types of books (and novels are certainly one of them) in order to meet industry standards. For some nonfiction books, four-color separation is merely an expensive option, but one well worth the cost if you can afford it. For example, I used the four-color separation process for the color photograph on the front cover of *How To Write, Publish & Sell Your Own Aviation Books*. Although this added roughly $800 to my printing costs, it was worth the extra expense because now you can see exactly what the cover of *Black Falcon* looks like in full-color.

This entire chapter might seem like a nightmare to you, but it really isn't that difficult to understand once you have your commercial artist and printer explain it all to you. *The Self-Publishing Manual* also explains in much greater detail most of the items I've mentioned in this chapter. (To order a copy of *The Self-Publishing Manual*, please refer to the *Writer's Book Mart Catalog* in the appendix section of *How To Write, Publish & Sell Your Own Aviation Books*.)

The processes described within this chapter are actually so simple, that I was able to design the cover of *How To Write, Publish & Sell Your Own Aviation Books* all by myself. I even made my own cover mechanical and overlay. That's right! I've learned a lot as a self-publisher and I'm proud of it!

CHAPTER THIRTEEN

PRINTING & BINDING

Printing and binding are the final two steps of the publishing process - the steps that will turn your camera-ready copy, cover mechanical, artwork and photographs into a printed and bound book.

Printing and binding are done by companies that specialize solely in the production of books. These companies are called short-run book printers and there are hundreds of them located all across the United States. I have found that the most economical quotes seem to come from firms that are located in the north-central region of the country (Michigan and surrounding states). Why is this so? Because the bulk of the paper manufacturing industry is located in this part of the country, making it a lot less expensive for book printers based in these states to purchase their paper supply and, in turn, allowing them to pass along their cost savings to their customers. You see, it's much more economical to ship finished books to distant places than it is to ship raw paper stock to those same places.

Before contracting out the printing and binding of your book you will need to get quotes from several short-run book printers. To get these quotes, simply prepare a Request For Quotation (RFQ) similar to the one you used for your typesetting quotes. Make as many copies as you need and send them to the printers you've chosen as possible candidates for the job. Keep in mind that the quotes you will be receiving are merely pre-publication estimates and that the final cost may vary slightly due to minor changes that occur during the production stage.

Before I describe how to write an RFQ for printing and binding quotes, let me first define what I mean by a short-run book printer. To begin with, these companies stay away from posters, brochures, newsletters and anything else that might detract from their specialty, which is printing and binding books. To do this, they have printing and binding machines that are set up specifically for the intricacies of book production, thereby eliminating the wasteful inefficiencies that are found at local, run-of-the-mill print shops, which are usually set up to print and bind a variety of items.

The result of this specialization is a great cost savings of thousands of dollars. In other words, small publishers can expect to get low-cost book production by using short-run book printers - and I do mean *low cost*.

The term short-run refers to the production of 5,000 copies or less. Some short-run book printers will produce as few as 200 copies, but 1,000 to 2,000 copies is more common and more economical for first-time self-publishers. Why? Because the fewer copies you produce, the higher your cost per copy will be (and vice versa). In fact, after the first thousand copies, the cost per copy comes way down. Don't make the mistake of producing too many copies, though, because that will cause you nothing but problems. I know a self-publisher who printed up 15,000 copies of his first book and got stuck with almost 14,000 of them in his garage after a year in print. If he had printed up 1,000 copies, he would have made a profit and then could have gone back for a second printing. Keep in mind that if you find your book is selling well, you can always go back for another printing - a much better situation than having thousands of unsold copies on hand.

Using *Black Falcon* as an example, I printed up 1,000 hardcover copies for the first printing, sold them in approximately two years, then went back for a second printing. This novel is now in its third printing, with a fourth printing planned for the future. *How To Get Published - Guaranteed* was started with a first printing of 1,500 copies in a trade paperback edition, while the book you are now reading was started with a first printing of 1,600 copies in a trade paperback edition. Are you getting the idea of how successful self-publishers work? They never print up too many books at any given time unless they know *positively* that they can sell that many books in a short period of time.

How can you find the names and addresses of short-run book printers? Well, *The Self-Publishing Manual* has an appendix that lists several dozen quality book printers. This is the reference source I've always used when sending out RFQs. In fact, the appendix alone is well worth the cost of *The Self-Publishing Manual*. By using the appendix, you can find a quality printer who specializes in the quantities, sizes and bindings that you want for your book. (To order a copy of *The Self-Publishing Manual*, please refer to the *Writer's Book Mart Catalog* in the appendix section of *How To Write, Publish & Sell Your Own Aviation Books*.)

After selecting several short-run book printers, your next step is to obtain printing and binding quotes from them by sending out RFQs. Figure 13-1 shows the RFQ I mailed out to twenty short-run book printers for the first printing of *Black Falcon*. As I received responses from the printers, I was quickly able to narrow my choices down to four companies (all of which were located in Michigan). The cost estimates from these four book printers were $4,300, $4,500, $4,600 and $4,900, which covered the cost of producing 1,000 hardcover copies with four-color dustjackets (the most expensive type of book cover), plus freight charges to South Florida. The other quotes I received ranged as high as $8,600 for the identical job - almost double the cost.

Because the four finalists had all given me quotes that

fell within the same price range, I had a difficult time making a final choice. After much thought, I decided to choose my book printer on the basis of how well each company answered the numerous questions I had about printing and binding - and I had plenty. You see, I was still a neophyte in the publishing trade at the time and needed lots of guidance.

The company that ended up getting my business was BookCrafters, but not because the other three companies were bad. In fact, just the opposite was true. From the sample books I received from each of the other printers, I knew perfectly well that each one of them could have done just as fine a job with *Black Falcon* as BookCrafters did. I later used this knowledge when it came time to print and bind *How To Get Published - Guaranteed*. Rather than sending out RFQs to dozens of book printers, I only sought quotes from the four finalists I had chosen during the selection process for *Black Falcon*. Figure 13-2 shows the RFQ I sent out for the first printing of *How To Get Published - Guaranteed*. The book printer that eventually landed that job was McNaughton & Gunn.

Now let's backtrack a bit and take a closer look at the responses I received from the twenty short-run book printers who had sent quotes for the first printing of *Black Falcon*. I already told you about the quotes I received from the four finalists, but what about the other sixteen printers? Why did I reject them? Well, ten had sent quotes that were either too high or had been sent in an unprofessional manner, three had sent letters stating that they needed more information before quoting the job, two had sent brochures without quotes, and one had sent an honest letter explaining that my RFQ did not fit into their efficiency range.

What did the last printer mean by efficiency range? Simply that each book printer has printing and binding machines that are set up to operate at different levels of efficiency - either by the size of the book, or number of copies

TRI-PACER PRESS, P.O. BOX 840111, Pembroke Pines, Florida 33084 • 305-432-9683

Publisher of Quality Action-Adventure Novels

REQUEST FOR QUOTATION — CAMERAWORK, BOOK PRINTING & BINDING

To: BookCrafters

Please quote your best price and delivery for camerawork and production of the following book from camera-ready copy:

SPECIFICATIONS
Name of book: (Black Falcon)
Type of book: (Fiction)
Total number of pages including front and back matter: (272)
Trim size: (5½ x 8½)
Copy: (Customer to provide camera-ready copy)
Cover: (Dustjacket: color separation from artwork, typeset mechanical, [1] B&W photo)
Paper: Inside: (70# vellum or similar high-bulk white book stock)
Dustjacket: (80# enamel cover stock)
Ink: Text: (Black throughout)
Dustjacket: (Four-color separation plus varnish coat)
Binding: (Hardcover, smyth sewn, cloth, name of book and author imprinted on spine)
Packaging: (In cartons)

QUOTE

Cover: Dustjacket: Quantity: (2000 + 100 extra):

Book:			
	First run of: (500):	(1000):	(2000):
	Later run of: (500):	(1000):	(2000):

Delivery charges to our address:

Delivery will be:_____working days from receipt of camera-ready copy and mechanical.

Terms:

Remarks:

Signed:________________________

Date:_____________

Any item in this RFQ takes precedence over any industry convention. The camera-ready copy, flats and all artwork shall be returned to customer on completion of job.

Please send a sample of your recent work plus a sample of your cover stock, book stock and a book planning guide (if available).

Figure 13-1

TRI-PACER PRESS, P.O. BOX 840111, Pembroke Pines, Florida 33084 • 305-432-9683

Publisher of Quality Action-Adventure Novels

REQUEST FOR QUOTATION – CAMERAWORK, BOOK PRINTING & BINDING

To: McNaughton & Gunn

Please quote your best price and delivery for camerawork and production of the following book from camera-ready copy:

SPECIFICATIONS
Name of book: How To Get Published – Guaranteed.
Type of book: Nonfiction.
Total number of pages including front and back matter: 288 (also quote +8, +16, +24, +32).
Trim size: 5-3/8" x 8-1/2" or 5-1/2" x 8-1/2". Camera-ready is now in the first trim size.
Copy: Customer to provide camera-ready copy and 10 color photos to be halftoned in B & W.
Cover: Customer to provide camera-ready mechanical and 1 black and white photo for halftone.
Paper: Text: 60# white with approximately 400 - 450 PPI or similar house book stock.
Cover: 10 pt. C1S white with lay flat film lamination.
Ink: Text: Black throughout.
Cover: Black ink plus one PMS color on white cover stock.
Binding: Perfect bound.
Packaging: Shrink wrap individually with clear plastic, then bulk in cartons.
Shipping: On pallet via truck freight.

Note: Need bluelines for text and cover.
Need price for extra covers: 200, 500, 1000, 1500, 2000, 2500.
Need reprint prices.

QUOTE
Please breakdown all costs. NO UNDERRUNS

First press run of 1500 copies:
Later press run of 1500 copies: 2000 copies: 3000 Copies:

Delivery charges to Florida:

Delivery will be:_____working days from receipt of camera-ready copy and mechanical.

Terms:

Remarks:

Signed:________________________Date:______________

Any item in this RFQ takes precedence over any industry convention. The camera-ready copy, mechanical and photographs shall be returned to customer on completion of job, as well as flats.

Please send a sample of similar perfect bound trade paperback using same paper and cover.

Figure 13-2

in a print run, or some other factor involved in book production. For instance, some might operate more efficiently with a print run of 500 copies, while others might operate more efficiently with a print run of 1,000 copies, or 2,000 copies, or even 5,000 copies. Some book printers might have machines that only bind paperback books, while others might only bind hardcover editions. In order to get a cost-efficient quote, the specifications listed on an RFQ must match the printer's capabilities. An analogy in the aviation industry would be the various types of airplanes used by different airlines. It would be foolish for a regional airline to operate a B-757 on a 300-mile route segment when an EMB-145 would be much more efficient. In the same manner, a major airline wouldn't even think of operating an EMB-145 on a coast-to-coast route from Florida to Califorina. Just think of how many refueling stops that would require! It's the same with printing companies - sometimes an RFQ does not match the efficiency range of their equipment.

Now that you have a bit of knowledge about short-run book printers and their different efficiency ranges, the next thing to learn is how to write an RFQ for printing and binding. For this discussion, we will continually be referring back to Figure 13-1 (the hardcover RFQ for *Black Falcon*) and Figure 13-2 (the paperback RFQ for *How To Get Published - Guaranteed*), plus looking at Figure 13-3 (the initial quotation from BookCrafters for the production of the first printing of *Black Falcon*), Figure 13-4 (the revised quotation from BookCrafters for the production of *Black Falcon*) and Figures 13-5 and 13-6 (the 2-page quote from McNaughton & Gunn for the production of the first printing of *How To Get Published - Guaranteed*).

Keep in mind that when I wrote the RFQ for *Black Falcon*, I was a novice in the areas of printing and binding, so I wasn't quite sure about the proper usage of certain terms. Fortunately, BookCrafters was able to decipher what I meant, and they corrected the terminology on their quotation sheets. By the time I wrote the RFQ for *How To Get Published - Guaranteed*, I had gained enough experience to use the correct terms myself. Because of this, the terminology used on the RFQ for my second book was similar to the terminology used by McNaughton & Gunn on their quotation sheets.

Let's start off our discussion by looking at Figure 13-1, the RFQ for *Black Falcon*. As you can see, I listed the title of the book (*Black Falcon*), the type of book (fiction), the approximate page count (272), and the trim size of the book (51/2" x 81/2"). These items are all self-explanatory, so no further discussion is necessary.

Next, I indicated that I would be providing the camera-ready copy, which let the printer know that the typesetting would be done elsewhere. If you now look at Figure 13-3, you will see that BookCrafters referred to my camera-ready copy as camera-ready repro's.

Going back to Figure 13-1, you can see that the dust-jacket instructions were that I would need a color separation from the artwork. I also indicated that I would be providing the mechanical and one black and white photograph. On the quotation from BookCrafters in Figure 13-3, they wrote that they would furnish the color separations, meaning that they understood about having to put my multi-colored artwork through the four-color separation process.

For the paper stock, I had listed 70# vellum or similar high-bulk white book stock for the text of *Black Falcon* and 80# enamel cover stock for the dustjackets. I didn't know anything about ordering paper at that time, but I had received a sample of vellum (a type of book paper) from another book printer and really liked it's texture, so that's what I asked for in my RFQ. BookCrafters didn't have 70# vellum in stock, but they offered me a similar paper called 60#

BookCrafters
140 Buchanan St./Post Office Box 370
Chelsea, Michigan 48118-0370/(313) 475-9145
Fax/(313) 475-7337

August 25, 1988
No. 02-108332 Rev. 1/047/H330-596001

Richard Cook
Tri-Pacer Press
P.O. Box 840111
Pembroke Pines, FL 33804

Re: BLACK FALCON

We are pleased to submit our specifications and prices:

QUANTITIES:	1,000 (no underruns) Add'l 100's
TRIM SIZE:	5-1/2 x 8-1/2 → Trimmed three sides (RC)
PAGES:	~~272~~ Change page count to 264 (RC)
COPY:	Text - camera ready repro's; without bleeds D/J - BookCrafters furnished color separations of 1 subject final size 5-3/4 x 8-3/4
PRESS:	Text - via offset lithography; black ink throughout Cover - via offset lithography; four color process with Duro Sheen
STOCK:	Text - 60# Lakewood White, 450 PPI D/J - 80# Enamel
BINDING:	Case bind with Kivar 5 cloth, Black, regular color & finish over .088" binders boards; 80# plain White end sheets; Smyth sew in 32's; rounded & backed; headbands and footbands; stamp spine with 1 impression imitation gold foil with copper dies - dies at add'l cost; wrap dust jackets
PACKING:	Bulk pack in single wall RSC cartons on pallets
SHIPPING:	FOB B/C dock, point of manufacturing
TERMS:	Credit to be arranged

Please adjust price down due to 8 less pages than quote. (RC)

PRICES:	1,000	Add'l 100's
	$4,280	$212

Approximate freight from Chelsea, MI to Pembroke Pines, FL 33084: $ 217 $ 15
Add'l for dustjacket color key $65

Add'l for blueline proofs of text: $123
Add'l for sample case: $40
Add'l for 200 add'l dust jackets: $14
Add'l for printed check copy: Postage only

Dust jackets (200 extra) are to be shipped to us as soon as possible for use as an advertising tool. (RC)

Camera ready boards and all artwork mechanicals, four color separations, photos and page negatives, (flats) shall be considered property of Tri-Pacer Press and will be shipped to Tri-Pacer Press upon completion of the job and payment in full.

Enclosed is a check for $2370.00 equal to 50% payment.
Time estimate per R. Senn dated 8/3/88 is 8 weeks.

These prices will be honored for 60 days from the above date, subject to price increases for materials which may occur prior to final shipment of the order. The price may be adjusted for changes, additions or alterations made by you in any specifications. Final prices will be sent to you after manufacturing has been initiated.

Orders based upon this estimate contain and acknowledge the acceptability of the terms, covenants and conditions on the face and reverse hereof, including all customs and usages of the trade and warranty disclaimers appearing herein.

Accepted by [signature] Title Editor-in-Chief Date 9/2/88

CUSTOMER COPY

Figure 13-3

BookCrafters
140 Buchanan St./Post Office Box 370
Chelsea, Michigan 48118-0370/(313) 475-9145

October 17, 1988
No. 02-108332 Rev. 1/047/596003
Job No. 17-77503
305/432-9683

Richard Cook
Tri-Pacer Press
P.O. Box 840111
Pembroke Pines, FL 33084

Re: BLACK FALCON

This is to confirm the prices for your order on the above title. Any changes made by you during production of the order will affect the final price. Other additional costs such as overrun quantities and shipping charges may also be added.

PRICES:		
	Base price per our Quote No. 02-108332, Rev. 1 dated 8/25/88 for 1,000 copies with 272 pages ---	$4,280.00
	Less for 8 less pages ----------------------------	(31.00)
	Add'l for add'l cover color prep ----------------	275.00
	Add'l for color key of dust jacket --------------	101.70
	Add'l for text bluelines -------------------------	118.80
	Add'l for stamped sample case -------------------	28.00
	Add'l for 200 extra dust jackets (four color plus 1 PMS color)--------------------------------------	21.00
	Add'l for double opaqing ------------------------	84.48
	New total	$4,877.98

Overrs - 2.095 — Freight - 217.00

$5094.98

Dies additional

Compositiond additional

Terms - 50% with order, 50% with shipment

FOB - Chelsea, MI

Thank you for your order. For order information, please call me at (313) 475-9145, extension 280.

Ed Blissick
Account Executive

EB/cc
cc: R. Sevin - (BookMasters 1-800-537-6727)

Figure 13-4

QUOTATION

M&G
McNaughton & Gunn, Inc.

960 Woodland Drive
P.O. BOX 10 Saline, Michigan 48176 313—429-5411

Contact: Dawn M. Schoeff
Ext. 265

RICHARD COOK
TRI PACER PRESS
PO BOX 840111
PEMBROKE PINES FL 33084

Date October 5, 1993
Quote #041724

Phone: 305-432-9683

TITLE:	H/T Get Published-Guaranteed
Quantity:	1,500; 2,000; 3,000
Page Count:	288
Trim Size:	5-1/2 x 8-1/2
Composition:	Camera ready line copy with 10 halftones on text. Camera ready color separated line mechanical on cover.
Press:	Black ink used throughout text. Black and 1 PMS color with film lam on covers 1, 4 & spine.
Stock:	60# Publishers Smooth 444 PPI on text. 10 PT. C1S on cover.
Binding:	Perfect bound.
Packing:	Individual shrink wrap and bulk pack in standard cartons on skids.
	See page 2 for pricing.

FREIGHT & POSTAGE—Billed at actual cost plus 5% handling charge or $5.00, whichever is greater unless shipped freight collect.

OVERRUNS OR UNDERRUNS—Overruns or underruns not to exceed 10% of the amount ordered shall constitute acceptable delivery and the excess or deficiency shall be charged or credited to the publisher at the overrun rate, not the unit rate.

Please sign and return the original copy of this quotation authorizing McNaughton & Gunn, Inc. to proceed with the manufacturing of this title. Note any changes in specifications and an amended quotation will be forwarded. See the reverse side for additional terms and conditions of this quotation.

Accepted by ______________________ Date ______________________

DP 5

Figure 13-5

QUOTATION

M&G
McNaughton & Gunn, Inc.

P.O. BOX 10 Saline, Michigan 48176 313—429-5411

Date October 5, 1993

Page 2 of quote #041724

TRI PACER PRESS

TITLE: H/T Get Published-Guaranteed

		1500 copies	2000 copies	3000 copies	Per add'l 100 copies
Cover:	Prep	74.00	74.00	74.00	-----
	Plate	43.00	43.00	43.00	-----
	Press	128.00	130.00	134.00	2.00
	Stock	79.00	96.00	131.00	4.00
	Coating	94.00	122.00	179.00	6.00
Text:	Prep	327.00	327.00	327.00	-----
	Plate	110.00	110.00	110.00	-----
	Press	495.00	619.00	868.00	26.00
	Stock	726.00	961.00	1,433.00	48.00
	Bind	975.00	1,168.00	1,500.00	35.00
	Total	3,051.00	3,650.00	4,799.00	121.00
	Appr.Freight	374.00	499.00	631.00	25.00

Alt. to Reprint

1,500 copies :	2,845.00
2,000 copies :	3,444.00
3,000 copies :	4,593.00
Per add'l 100's :	121.00

Add'l for text blues : 135.36
Add'l for cover blue : 10.00 each.
Add'l for extra covers :

200 :	30.00
500 :	75.00
1000 :	150.00
1500 :	225.00
2000 :	300.00
2500 :	375.00

FOB: Saline, Mi.

Terms: Subject to credit review. Schedule subject to credit approval.
Quote valid for 45 days, excluding material increases.

Dawn M. Schoeff
Sales Marketer

FREIGHT & POSTAGE—Billed at actual cost plus 5% handling charge or $5.00, whichever is greater unless shipped freight collect.

OVERRUNS OR UNDERRUNS—Overruns or underruns not to exceed 10% of the amount ordered shall constitute acceptable delivery and the excess or deficiency shall be charged or credited to the publisher at the overrun rate, not the unit rate.

Please sign and return the original copy of this quotation authorizing McNaughton & Gunn, Inc. to proceed with the manufacturing of this title. Note any changes in specifications and an amended quotation will be forwarded. See the reverse side for additional terms and conditions of this quotation.

Accepted by ______________________ Date ______________________

DP 5

Figure 13-6

Lakewood White that had a bulking factor of 450 PPI (pages per inch). For the dustjackets, they were able to supply me with the 80# enamel I had asked for.

This might all seem confusing to you right now, but it really isn't that difficult to understand. Paper is paper, but paper comes in different types and different qualities. Basically, any printer can order any kind of paper you might want, whether it be coated book stock, fancy textured book stock, acid-free book stock, recycled book stock or something altogether different. The important point to keep in mind is that if you request a paper that your printer does not normally keep in stock, your production costs will increase dramatically. Why? Because most book printers purchase paper by the truck-load (and sometimes even by the train-load), which gives them a lower cost for the over-all shipment of paper. This translates into lower prices for their customers. This paper is called house paper. An analogy can be made to the cost of a glass of house wine in a restaurant compared to the cost of a rare or vintage wine. If a publisher requests a special type of paper other than the house paper, the cost of acquiring that paper will be very high because not only does the printer have to order it in a small quantity, but he also must order it in a large enough quantity to have some extra stock on-hand in case mistakes are made during the production of the book. Naturally, the publisher pays for it all.[50] For most self-publishers, house paper works fine.

As for the bulking factor of paper, this is just a way of referring to the thickness of a particular type of paper. For instance, BookCrafters offered me their house paper, which had a bulking factor of 450 PPI. This meant 450 pages per inch - a page being one side of a sheet of paper.[51] In other words, a 225-page book would be half an inch thick when bound.

On my RFQ in Figure 13-1, I was correct in describing

the paper for the dustjackets as being enamel, but wrong in calling it cover stock. Cover stock is the cardboard-like paper used for softcover books, as opposed to the thinner paper used for the dustjackets of hardcover books. Enamel is measured in pounds (such as the 80# enamel BookCrafters used for *Black Falcon*), while cover stock is measured in points (such as the 10 pt. C1S cover stock McNaughton & Gunn used for *How To Get Published - Guaranteed*, which is also the same type of cover stock used for *How To Write, Publish & Sell Your Own Aviation Books*).[52] If you look at the RFQ in Figure 13-2, you will see that I ordered the cover stock for *How To Get Published - Guaranteed* with a lay flat film lamination. This lamination gives the book cover a shiny appearance and also prevents it from curling up due to moisture in the air (a real problem with softcover books).

For the ink specifications, I indicated that I wanted black ink throughout the text for both *Black Falcon* and *How To Get Published - Guaranteed*. For the cover ink specifications, I indicated four-color separation plus a varnish coat for *Black Falcon*, and black plus one PMS color for *How To Get Published - Guaranteed*. On the quotations from BookCrafters and McNaughton & Gunn, these were written as four color process with Duro Sheen for *Black Falcon*, and black plus one PMS color with film lamination for *How To Get Published - Guaranteed*. The quote for *How To Get Published - Guaranteed* was later changed because I decided to use the four-color separation process on the front cover photograph rather than printing it in black and white. This added $770 to the total cost of printing *How To Get Published - Guaranteed*.

Because I ordered *Black Falcon* as a hardcover book and *How To Get Published - Guaranteed* as a trade paperback, the binding specifications were quite different for each book. The fact of the matter is that it's much easier to produce a paperback book than it is to produce a hardcover book - thus, the specifications are much simpler for a paperback book. Because of this, let's first look at the binding specifications for *How To Get Published - Guaranteed,* which can be seen in the RFQ in Figure 13-2. As you can see, I merely stated that I wanted the book perfect bound. That explained it all to McNaughton & Gunn. In the book manufacturing industry, perfect bound means a softcover book that has the cover glued to the spine of the text. Now look at the quotation I received from McNaughton & Gunn in Figure 13-5 regarding the binding for *How To Get Published - Guaranteed.* Just like on my RFQ, their quote simply stated that the book was to be perfect bound. The only term you will need to know in regards to paperback books is notch binding. This is nothing more than grooves being cut into the spine of the text, which allows a bit of extra adhesion area for the glue on a perfect bound book. By the way, when I mention trade paperback books, I'm referring to full-sized softcover books (usually 51/2" x 81/2" and larger) that sell side-by-side with hardcover books on bookstore shelves, as opposed to mass-market paperback books, which are smaller in size (usually 4" x 7") and sell on racks in airport terminals, grocery stores, drug stores, and convenience stores.

Now let's take a look at the hardcover binding I ordered for *Black Falcon.* The RFQ in Figure 13-1 shows that I asked for the binding to be hardcover, smyth sewn and cloth. I also requested that the title of the book and the author's name be imprinted on the spine of the binding. Well, as I already told you, I was a novice when I typed up the RFQ for *Black Falcon* and my specifications left out a lot of information. The people at BookCrafters, however, were somehow able to figure out what I really wanted for the binding of my novel. Looking at their quotation in Figure 13-3, you can see that they listed the binding as being case bound with black Kivar 5 cloth in regular color and finish over .088" binders boards, 80# plain white end sheets, smyth sewn in 32's, rounded and backed, headbands and footbands, stamped spine with one impression of imitation gold foil with copper dies, and dustjackets wrapped. What does this all mean? Well, let's start by examining the specifications I had listed on my RFQ and then discuss what BookCrafters wrote on their quotation sheet.

First of all, I had asked for *Black Falcon* to be bound in a hardcover binding using the smyth sewn method. Smyth sewn means that all of the pages of a book are actually sewn together using threads. I'm sure you've seen these threads while reading hardcover books, but have you ever wondered what their purpose was or what they were called? Now you know.

Next, I had asked for a cloth covering to be placed over the hardcover binding. I knew that coverings could be ordered in either paper or cloth (with paper being the less expensive), but I wanted my book to look more professional so I went with the cloth. I'm sure you've seen both types of covering during your years of reading books.

My last specification for the binding was to have the title of the book and the author's name imprinted on the spine of the cloth covering. This step is not necessary, but it's usually done anyway for most hardcover books because dustjackets get old after awhile, sometimes even getting torn or misplaced, and the imprinting allows people to still know what book it is without having to look on the title page. Libraries especially like this feature on books.

Now let's compare the specifications listed on my RFQ to the ones listed on the quotation sheet sent to me by BookCrafters. First of all, BookCrafters did not use the term hardcover binding anywhere on their quotation sheet. Instead, they used the term case binding. Secondly, because they offered cloth coverings in a variety of styles and grades, they chose their house cloth (the same principle as house paper) for *Black Falcon*. This house cloth was called Kivar 5 and the color chosen for the cloth was black (which worked out great because it blended in with the mostly-black dustjackets of *Black Falcon*).

BookCrafters also specified that .088" binders boards would be used for my novel. Binders boards are the heavy cardboard sheets that give a hardcover book its hardness. These boards come in various thicknesses, depending on the size of the book. For a 51/2" x 81/2" book with approximately 250 pages (the size of *Black Falcon*), .088" thick cardboard is the standard size used in order to give the book strength and heft.

BookCrafters also specified that 80# plain white end sheets would be used for *Black Falcon*. I had totally forgotten about end sheets, which are the heavy pieces of paper that reinforce the text securely to the binders boards.

Another specification brought out by BookCrafters was that *Black Falcon* would be smyth sewn in 32's, which meant that every 32 pages of text would be individually sewn together into units called signatures, then all the signatures would be sewn together into the body of the text, thereby completing the smyth sewn process.

BookCrafters also stated that the binding would be rounded and backed, which meant that the spine would have a convex (curved) look to it and would also be wider so that the two binders boards for the front and back covers would rest against it as an additional means of support.

One more specification listed by BookCrafters was that the binding would have headbands and footbands. Headbands are brightly colored scarfs worn by sweaty printshop workers. (Just kidding. I couldn't resist throwing in a bit of publishing humor. I know what you're thinking, though. I should stick to my writing and stay away from comedy.) Anyway, headbands and footbands are those little cloth inserts at the top and bottom of the spine on hardcover books.

I'm sure you've seen them before, but never paid any attention to them. These pieces of cloth are not necessary, but they give hardcover books a finished appearance.

And finally, we close out the specifications for *Black Falcon* with the imprinting of the title and author's name on the spine of the book. This process uses metal dies to create the lettering. The cost of these dies is additional, but the expense is minimal for the added value it gives to the books, especially if you're planning on selling to libraries.

Now let's take another look at Figure 13-3 to go over the additional notes that BookCrafters had written on their quotation sheet. The references to packing, shipping, credit terms and freight charges are all self-explanatory. They mean what they are. On the other hand, the note in reference to the additional 200 dustjackets might not be self-explanatory. After all, why would I want extra dustjackets? For promotional purposes - that's why! And you should, too! You see, once the printing press is set up and rolling, it's no trouble at all to print up extra book covers at very little extra cost. With these 200 extra dustjackets, I was able to send a dustjacket with each press release when *Black Falcon* finally came out in print - sort of like an actual size promotional poster of the book. In fact, these extra dustjackets were so useful and so inexpensive that I had 1,500 extra dustjackets printed up during the second printing of *Black Falcon*. This only added $222 to my total production cost. Why did I have so many printed up? Well, I had already proven that I could sell out the first printing of *Black Falcon* and I knew I'd eventually be going back for a third and fourth printing of the book, so I figured that as long as the printing press was rolling I'd print up enough dustjackets during the second printing to cover my next two printings as well, thereby saving me about $1,000 in printing costs for each of those two additional printings. In other words, for my $222 up front investment, I saved $2,000 down the road, plus it gave me plenty of extra dustjackets to use for promotional purposes - enough to last me for many years.

Another additional note that BookCrafters mentioned on their quotation sheet had to do with blueline proofs. I know for sure that bluelines are not self-explanatory, so I'll try to describe them the best I can. To begin with, blueline proofs are optional. You don't have to order them if you don't want to spend the extra money. However, by not ordering bluelines you would be canceling out the only means you have of assuring that your book will be printed correctly. You see, book printers make mistakes, too, just like everybody else. After all, they are only human. For example, it's very possible to have a page printed upside down or out of sequence, or for a photograph to be printed on the wrong page. Bluelines are your means of making sure that this does not happen to your book. Basically, bluelines are pre-press page proofs that are created after the printer has set up all the pages and photographs, but before the actual printing is done. Therefore, if the printer has made any mistakes, you can catch them on the blueline proofs. That way, the printer will be able to correct the mistakes before they end up in every copy of your book. However, book printers will not send you blueline proofs unless you order them in advance, so you'd be foolish not to order them. The quote for the bluelines for *Black Falcon* came to $123 - not a large sum of money to assure that all the pages were printed correctly.

Now let's discuss the final amount I paid for the first printing of *Black Falcon*. Figure 13-3 shows that the original quote came to $4,280 for printing and binding, $123 for blueline proofs, $14 for 200 extra dustjackets and $217 for freight charges to Florida - a total of $4,634 for 1,000 hardcover copies of *Black Falcon*. This was not what I paid, however. The revised quote shown in Figure 13-4 shows that a deduction of $31 was made because the final page count was eight pages less than originally planned. Figure 13-4 also shows a reduction of the blueline charges from $123 down to $118 - another result of the lower page count. The revised quote in Figure 13-4 also shows a cost increase from $14 up to $21 for the 200 extra dustjackets - a result of additional color preparations that were necessary during the production of the dustjackets. Figure 13-4 also shows additional charges that were added to the total bill for each of the following items: $275 for color preparations that were not taken into consideration on the original quote (knockouts, drop outs, reverses, etc.), $101 for a color key of the dustjacket (a color proof of the four-color separation artwork), $28 for a sample stamped case (a proof copy of the hardcover binding) and $84 for double opaquing the camera-ready copy (which was necessary to eliminate the shadows caused by the sloppy typesetting and paste-up job).

With all of these deductions, reductions, increases and additions, the final printing and binding charges came to $5,095 for 1,000 copies of a 257-page book, including freight charges to Florida. By adding in the $2,640 I had paid for typesetting and paste-up work, plus miscellaneous costs of approximately $150, the total production charges for *Black Falcon* added up to $7,885, which I'll round off to $7,900 (or $7.90 per copy for a book that had a retail price of $17.95 for the first printing).

You might be thinking that $7,900 is a lot of money to spend for a self-published book, but please take note here that it doesn't have to cost this much to self-publish a book! *Black Falcon* was my first attempt at self-publishing and I made a few costly mistakes. I learned from my mistakes, though, and was able to lower the production costs of my second book by several thousand dollars. Figure 13-6 shows the quote I received from McNaughton & Gunn for printing and binding 1,500 copies of *How To Get Published - Guaranteed* in a trade paperback edition, plus bluelines, 200 extra covers and freight charges to Florida. The quote came to only $3,600 for printing and binding. This quote, however, was for my original estimate of a 288-page book and a black and white photograph on the front cover. The final page count of 328 and my decision to print to front cover photograph in color instead of black and white raised my printing and binding costs to $4,600 for 1,500 copies. Adding in the $2,000 I had already paid for typesetting, the cost of producing my second book came to $6,600 for 1,500 copies (or $4.40 per copy for a book that had a retail price of $18.95 per copy for the first printing).

Do you think it was worth having spent $6,600 in order to have gotten $28,425 worth of books (1,500 x $18.95) or are you still not convinced? How about if I had known how to do my own typesetting on a personal computer when I had written *How To Get Published - Guaranteed*? Then my total production costs would have been only $4,600 for 1,500 copies (or $3.07 per copy)! If you are still not convinced at this point, then self-publishing is definitely *not* for you. On the other hand, if this seems like something you'd like to attempt, then I've got an added bonus to tell you about. It's called reprinting. Once a book sells out of its first printing, it is less expensive to go back to press for reprintings. Why? Because the pre-press preparations have already been completed and the film negatives are ready to go. This means that reprintings involve less work for the printer, thus lowering the production costs. Reprintings are where self-publishers really start making decent profits if they do their marketing correctly. Hey, I'm not kidding you here. As long as you can write a good book, the potential is out there to make huge profits through self-publishing.

Now for the exciting news. If you can get to this point with *your* book, you will then be a published author. Never again in your life will you be referred to as an unpublished author. Think about this for a moment - the glory and prestige of being published! It's definitely something to be proud of, especially if you can do it all on your own. And if you make a few mistakes along the way, it's no big deal. For example, I made mistakes in regards to the publishing of *Black Falcon* (like the botched typesetting job, which fortunately my printer was able to fix for a mere $84), but I learned from my mistakes. All in all, I'd say that 95% of the decisions I've made in regards to self-publishing have been correct - and that's not bad for a blue-collar worker.

I feel proud of my accomplishments, especially since I've proven that a novice can successfully self-publish both fiction and nonfiction if enough energy and ingenuity are put into the projects. I'm also proud of the fact that I did my first two books without the aid of a computer. Sure, it would have been easier and less expensive with the help of modern technology (like I discovered while writing and publishing *How To Write, Publish & Sell Your Own Aviation Books)*, but I've proven to all of you out there who are computer illiterate that you don't need a computer to write and self-publish your own books. All you need is a pen, lots of paper and a good typewriter (plus a little extra cash to pay for the typesetting of the book).

The success of Tri-Pacer Press was guaranteed from the very beginning because I was the one who was in charge of my own destiny. Through the use of my own publishing company, I have achieved the goal of guaranteed publication for all of my books - past, present and future. I've told you a great deal of information about how you can achieve this goal for yourself. The choice is yours. *You have to be the one who guarantees the publication of your book!* Why? Because nobody else will do it for you!

Although I had a lot of fun writing *How To Write, Publish & Sell Your Own Aviation Books*, it took many years of research and a lot of personal experience to actually get the information needed to write this book. But guess what? It was all worth it - every early morning, every late night, every missed deadline, every everything! And I did it all for people like you. Why? Because I was once a struggling, unpublished aviation author myself (with a string of eighty-four rejections for *Black Falcon*). Then I found out about self-publishing. By writing *How To Write, Publish & Sell Your Own Aviation Books,* I have saved you years of intense research and heartache. Hopefully, my book will help guide you onto the road leading towards guaranteed publication.

But guess what? I'm not finished yet! There's a whole lot more that I still need to tell you about - a whole lot more! The next few chapters will teach you how to set up a pricing policy and discount schedule, how to handle collections and record-keeping and, most importantly, how to promote and market your books. So read on!

(Note: The following paragraph was added to the end of this chapter two weeks before the book went to press so that I could have the opportunity to show you how I selected a book printer for *How To Write, Publish & Sell Your Own Aviation Books.*)

One January 19, 1998, I sent out RFQs to twenty book printers. Within a month I had received quotes from twelve of those printers, ranging in cost from a low of $5,051 to a high of $7,952 for 1,600 trade paperback copies, plus 500 extra covers. I had also asked the printers to send sample copies of books they had printed so that I could check the quality of their work. The low-bidder, United Graphics, had sent several quality samples, so I called the publishers of those books to see if they were pleased with this printer (which they were). So I contacted Louis Segovia, the Sales Representative for United Graphics (1230 S. 57th Avenue, Cicero, IL, 60804, 708-780-7728) to arrange to have this book printed at their plant in Mattoon, Illinois. Once again I chose a book printer located in the north-central region of the country - this time at a cost of $3.16 per copy.

CHAPTER FOURTEEN

PRICING, COLLECTIONS & RECORD-KEEPING

As a publisher, you must decide on a retail price for each of your books. The retail price is the amount of money that will be collected from retail customers (usually these are the people who buy the books for their own enjoyment or as gifts for other people). The retail price of your book should be high enough to cover the cost of producing the book, plus your business expenses, plus a profit for all of your efforts. A self-publisher might not earn a large profit on the first printing of a book, but he or she should at least set their retail price high enough to afford a second printing, which is when a good profit will most likely kick in.

A standard rule-of-thumb for setting the retail price of a book is that it should equal at least four times the cost of printing and binding the book, but can go as high as eight times the cost of producing the book if thousands of copies are printed. Sometimes this rule-of-thumb is not possible. A good example is the first printing of *Black Falcon*. The cost of printing and binding this hardcover novel came to $5.10 per copy (excluding typesetting and paste-up costs, which were a one-time expense). Multiplying $5.10 by four gave me a retail price of $20.40, which was too high for a 257-page novel in 1988. However, by multiplying $5.10 by three-and-a-half, I came up with a figure of $17.85, which was then adjusted up to $17.95 to meet industry pricing standards of ending prices with .95 (such as $7.95, $14.95, $19.95, $24.95). This gave me a fair-market retail price for a 257-page hardcover novel in 1988. The second printing of *Black Falcon* in 1991 had a price increase to $18.95 due to inflation, and the third printing of this book in 1996 had a price increase to $20.95 (which means it is now set between four to five times the cost of production - well within the standard rule-of-thumb for retail pricing).

With *How To Get Published - Guaranteed*, I was able to multiply $3.07 (the cost of printing and binding the book) by six, which gave me a figure of $18.42, which was then adjusted up to $18.95 to meet industry pricing standards. This gave me a fair-market retail price for a 328-page trade paperback how-to book in 1994. The price for the second printing of this book will be increased to $21.95 when it comes off the press.

As a publisher, you should realize that not all of your books will be sold at their full retail price. Many copies will be sold to other businesses (called retailers) who, in turn, will sell your books to their own customers. These retailers expect to earn a fair profit from the sales of your books, therefore, they expect to buy the books from you at a reduced price. This is called discounting.

Until 1994, Tri-Pacer Press used a sliding discount schedule, meaning that the discount was 25% off the retail price if a retailer bought one or two books, 33% off if a retailer bought three or four books, and 40% off if a retailer bought five or more books. In 1994, however, I decided to simplify my discount schedule by offering an across-the-board 40% off for orders as small as one copy. This really increased the number of special orders I started receiving from large chain bookstores, such as Barnes & Noble. In fact, a single-copy order over the telephone from the Barnes & Noble bookstore in Schaumburg, Illinois, just outside of Chicago, lead to a speaking engagement at that store, which in turn lead to the sale of a full case of books (44 copies).

Besides retailers, you also might be selling your books

to businesses known as wholesalers and distributors. The difference between these two types of firms is that wholesalers only fulfill orders when bookstores and libraries come to them, whereas distributors actually send sales representatives out to bookstores and libraries. Distributors, therefore, get a very large discount - usually 65% - and can also bill you for such expenses as special promotions, advertising and catalog listings. Wholesalers, on the other hand, usually get a discount anywhere between 50% to 60%. I personally will not sell to anyone who demands more than a 55% discount (unless, of course, they want to buy an extremely large quantity of books on a non-returnable basis).

Many book wholesalers and distributors are slow in paying their bills - sometimes taking up to six months before sending checks to publishers. They are also infamous for returning lots of books. Because of all these factors, a lot of self-publishers refuse to do business with wholesalers and distributors. I do some business with wholesalers, but stay clear of distributors after having had a bad contractual relationship with one of these firms. In all honesty, I prefer dealing directly with retailers, but sometimes I'm obligated to deal with wholesalers in order to sell to certain retailers. *The Self-Publishing Manual* has an entire chapter on how to deal successfully with wholesalers and distributors. If you plan on selling books to these companies, then I suggest you read *The Self-Publishing Manual* thoroughly, otherwise you can end up losing a lot of money in the long run. (To order a copy of *The Self-Publishing Manual*, please refer to the *Writer's Book Mart Catalog* in the appendix section of *How To Write, Publish & Sell Your Own Aviation Books*.)

Once you start selling your books you will become involved with collections, which is nothing more than the process of receiving payment for books that are sold. For most self-publishers, this usually involves payment by cash, check or money order. Retail customers are expected to pay for their books at the time of the sale. Businesses (such as retailers, wholesalers and distributors), on the other hand, are usually granted special payment terms. The terms offered by Tri-Pacer Press are *Net 30 Days From The Date Of Invoice*. This means that they have thirty days to pay me. Of course, not every business pays on time. For example, most book wholesalers and distributors tend to pay between three to six months after the invoice date. I would estimate that in all my years of operating Tri-Pacer Press, only about 2% of my collections have been late. That's a good record for a small publisher.

All collections, whether retail or discounted, are recorded on invoices. Figure 14-1 shows an invoice from a retail sale that occurred on December 6, 1988. A customer from North Miami Beach, Florida, ordered one copy of *Black Falcon* via mail-order. The receipt shows a total payment of $20.53, which included $17.95 for the book (the retail price of the first printing), $1.08 for Florida sales tax and $1.50 to cover shipping (which has since gone up due

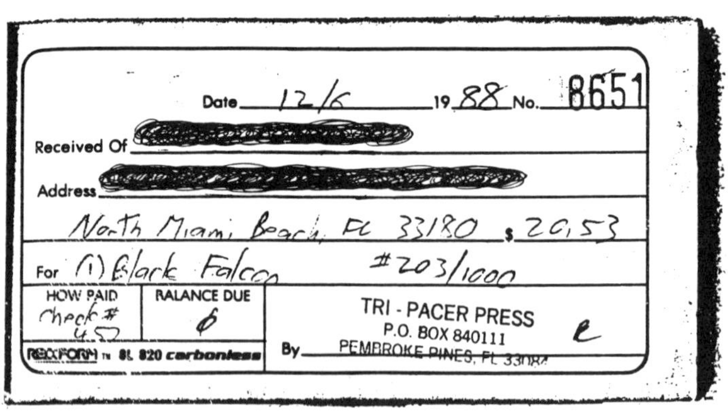
Date 12/6 19 88 No. 8651
Received Of
Address
North Miami Beach, FL 33180 $ 20.53
For (1) Black Falcon #203/1000
HOW PAID: Check # 452
BALANCE DUE: ∅
TRI - PACER PRESS
P.O. BOX 840111
PEMBROKE PINES, FL 33084
By

Figure 14-1

to inflation). The invoice came from a book of Rediform carbonless receipts. These receipt books can be purchased at most office supply stores. Upon completion of an order, the original invoice goes to the customer, while the carbonless copy stays with Tri-Pacer Press as a permanent record of the sale.

Invoicing procedures for commercial sales are only slightly more complicated than for retail sales. Again, Rediform carbonless receipt books are used for invoicing purposes. Figure 14-2 shows an invoice from a commercial sale that occurred on July 14, 1989. A retail outlet in Fort Lauderdale, Florida, ordered six copies of *Black Falcon*. Notice how I subtracted the discounted amount from the total retail price to come up with a sale price of $64.62 (which had a due date of August 14, 1989 because of my Net 30 Days payment policy). Also notice that I did not charge this customer for shipping. This was because I delivered the books personally on my way to work - thereby saving the customer some money. I only do this for local business accounts. All other commercial sales have shipping charges added into the bill.

Another point to notice in Figure 14-2 is that no sales tax was charged to this customer. That's because this customer had purchased the books to resell to his own customers (known as resale), at which time Florida sales tax would be collected. In other words, this was a tax-exempt sale. In order for me to legally sell my books in Florida without collecting Florida sales tax, I have to get a signed Blanket Certificate Of Resale from my Florida-based commercial customers. This law is in effect in every state that collects sales tax. Figure 14-3 shows a copy of a signed Blanket Certificate Of Resale. These 3" x 5" cards are sold at most office supply stores.

Now go back to Figure 14-2 and notice the statement in the middle of the invoice that reads *Pd. Check #037920*

7142

TRI - PACER PRESS
P.O. BOX 840111
PEMBROKE PINES, FL 33084

CUSTOMER'S ORDER NO.	DEPT.	DATE
6484		7/14/89

NAME [redacted]

ADDRESS [redacted] Ft. Lauderdale, FL 33316

SOLD BY	CASH	C.O.D.	CHARGE	ON ACCT.	MDSE RETD.	PAID OUT
R				X		

QUAN		DESCRIPTION	PRICE	AMOUNT	
6	1	Black Falcon	17.95	107	70
	2				
	3	Discount (-40%)		<43	08>
	4				
	5				
	6				
	7				
	8				
	9	(#'s 607-612)			
	10	pd. check #037920 8/30/89			
	11				
	12				
	13				
	14				
	15				
	16				
	17				
	18	Net 30 Days		64	62

REC'D BY

REDIFORM 5L320/01320

KEEP THIS SLIP FOR REFERENCE

Figure 14-2

Purchaser [redacted]
Address [redacted] Miami Springs, FL 33166 RULE 39
FLORIDA DEPARTMENT OF REVENUE, SALES TAX DIVISION
BLANKET CERTIFICATE OF RESALE
This is to certify that all material, merchandise, or goods purchased by the undersigned from the
Tri-Pacer Press
after 2/15/89 (date) *is purchased for the following purpose:*
☒ Resale as tangible personal property.
☐ To be incorporated as a material or part of other tangible personal property to be produced for sale by manufacturing, assembling, processing or refining.
☐ To be exported for sale, use, or consumption outside the continental limits of the United States.
☐ Other:
This certificate shall be considered a part of each order which we shall give provided such order contains our certificate number. This certificate is to continue in force until revoked
CERTIF. NUMBER 23-11-20579-27 BY
SEMINOLE R - 39

Figure 14-3

8/30/89. This was the date I received payment for the sale. This payment was late because it didn't arrive within the specified time period. As I've already mentioned, only about 2% of my sales have had late payments. Although this is a good record, late payments are late payments and they must be dealt with head-on. You see, late payments are a hassle because they waste valuable time and energy to collect the money - energy that could be used for more productive activities. Figure 14-4 shows the standard form letter I mail to commercial customers when their invoices become past due (meaning that the payments are late). I also enclose a photocopy of each invoice involved. This method of collecting late payments seems to work fairly well, although sometimes the payments still take a long time to arrive even after the letter is sent. The example shown in Figure 14-4 is a good case in point. The invoice for the sale was dated November 20, 1991, payment was due by the 20th of December, the late payment letter was sent out on January 6, 1992, and I finally received payment on the 4th of March. I have

TRI-PACER PRESS, P.O. Box 840111, Pembroke Pines, Florida 33084

Publisher of Quality Action-Adventure Novels

January 6, 1992

Boca Raton, FL 33432

Attention: Accounts Payable Department

A recent review of your account indicates an outstanding amount of $113.22 from our invoice #7282 dated 11/20/91.

Our terms are Net (30) Days from date of invoice. Please review your records and send us a check promptly for a total of $113.22. If payment is not received within (10) days, interest will be charged at (1½%) per month.

Thank you in advance for your help with this matter.

Sincerely,

TRI-PACER PRESS

Richard Cook
Editor-in-Chief
(305)432-9683

ENC: Photocopy of Invoice

Figure 14-4

only *not been paid* a few times - the major instance being the loss of $125 due to a book wholesaler filing for bankruptcy before paying me for the books they received.

In order for a self-publishing venture to survive as a business, accurate and detailed records must be kept for all financial matters. Record-keeping is a repetitive task that seems very time consuming at first, but soon becomes habit-forming and done without much effort. The primary purpose of record-keeping is to show the health and wealth of the business - total sales, total expenses, taxes owed, and so on and so forth. Without proper record-keeping, it would be impossible to keep track of a business - even a small business like Tri-Pacer Press.

Because I didn't particularly enjoy financial paperwork when I first started out as a self-publisher (and still don't), I devised a simple record-keeping system that worked quite well during the first five years I operated Tri-Pacer Press (when I only had one book on the market). When I expanded my business during its sixth year of operation, that simple system no longer worked for me, which caused Tri-Pacer Press to switch to a much more complicated record-keeping system - too complicated, in fact, to describe in this book because it involves more than half-a-dozen products (including fees from consulting jobs, speaking engagements and books by other publishers). For self-publishers who are just starting out, however, my original system will work just fine. If the time comes for you to expand your publishing company, then you should go seek professional advice from a Certified Public Accountant.

Basically, my original record-keeping system involved the use of three separate ledger books (expense ledger, sales ledger and vehicle mileage ledger), plus a series of file folders in which I kept important documents (bank statements, tax reports, purchase orders and expense receipts).

The first ledger in my original system was the expense

ledger, which showed a complete listing of business-related expenses in a twelve-column ledger book. These books are sold at most office supply stores. Figures 14-5 and 14-6 show one of the double-wide pages from my original expense ledger. Notice that each entry was listed with the date and description of the expense, plus the amount of the expense. The eleven categories that I used in my original system were: Postage, Office Supplies, Resource Books, Freelance Help, Book Production, Advertising, Food/Other, Travel/Lodging, Miscellaneous, Exhibit Fees, Dues/Fees.

Although *Black Falcon* wasn't published until October 1988, I was legally allowed to began recording my expenses towards the end of 1987 when I originally set out to start Tri-Pacer Press. Every business-related dollar that has been spent since that time has been used as a tax deduction on my annual IRS income tax form. On top of that, because Tri-Pacer Press has been operating out of my home (as both an office and warehouse), I've been allowed to legally deduct a certain percentage of my household-related expenses (mortgage, insurance, utilities, etc.) as part of my business-related expenses. If you decide to start your own publishing company, then I suggest that you seek professional advice from a Certified Public Accountant to find out what percentage (if any) of household-related expenses you can deduct for having an in-home business. Basically, the rule-of-thumb is that the tax deduction is equal to the percentage of the square footage of the home that is used exclusively for business purposes. Also, an entire room must be used solely for business purposes. In other words, you cannot set up a desk in your living room and call it a home office, thereby claiming the square footage of the living room as a tax deduction. For example, my entire den is used strictly as a combination office and warehouse for Tri-Pacer Press. The size of the den amounts to 10% of the square footage of my home - hence, I've been allowed to deduct 10% of my household expenses as part of running my business.

Now let's take a closer look at each of the eleven categories listed in the expense ledger. First of all, how did I arrive at these eleven categories? The simple truth is that they came to me via trial and error. I originally thought up the first eight categories, figuring that they would be the most likely expenses I would encounter as a self-publisher. Then I added the Miscellaneous category as sort of a catch-all for any unforeseen expenses that might come up while doing business. Finally, as I started doing more trade shows and special events, I had to add the Exhibit Fees category. At the same time I also added the Dues/Fees category. All eleven categories are simple enough for any self-publisher to work with. The following paragraphs are brief descriptions of each category listed in my original expense ledger.

Postage: Includes postage stamps, priority mail, certified mail, registered mail, return receipt charges, shipping insurance charges, UPS shipping charges, Federal Express shipping charges and any other expenses related to the mailing or shipping of letters, documents, books and packages. While I'm on the subject of postage, you should know that the U.S. Postal Service provides discounted postal rates for the shipment of books and manuscripts. When you go to the post office, ask for a free Fourth Class Book Rate postage chart. You will save your customers a lot of money on their shipping costs by using these special rates. Just remember to mark the outside of your shipping envelopes with a black or red ink stamper that indicates *Special Fourth Class Rate, Contents: Books-Manuscripts, Return Postage Guaranteed.* Figure 14-7 shows the stamper that I use on all of my book shipments. You can order these stampers at most office supply stores.

Office Supplies: Includes file folders, blank paper, pens, letterhead, envelopes, shipping tape, paper clips, staples, ink stampers, ink pads, typewriter supplies, computer supplies

Expenses

1988	#	Description	Postage	Office Supplies	Resource Books	Freelance Help
5/23	1	Membership (1) Year - World Airline Historical Society - check				
5/24	2	Stamps	30.00			
5/30	3	The Aviator Catalog - check				
5/30	4	COSMEP (1) year membership - check				
5/30	5	Small Press magazine (1) year sub. - check			25.95	
6/1	6	Bank Service Charge				
6/18	7	Typesetting (1/3 due) paid by check				
7/1	8	Bank Service Charge				
7/14	9	Vault paid by check				
7/14	10	Jacket Stats				
7/20	11	Laminate Display Ad.				
7/22	12	Balance of artists fee for cover art paid by check				50.00
7/25	13	Stamps	1.42			
7/26	14	Balance of Typesetting (1/3) paid by check				
7/26	15	Jacket Stats				
8/13	16	Keys Trip - Lodging, Tolls 8/11-8/13				
8/13	17	Food on Key's Trip 8/11-8/13				
8/16	18	Pasteup of Book paid by check				
8/17	19	3 Folders for typesetting copies		4.62		
8/17	20	Stat of jacket cover (from stat board)				
8/27	21	Stamps	26.85			
8/27	22	White coverup tape for pasteup		2.37		
8/27	23	Masonite Hardboard for artwork				
8/30	24	Laminate Jacket Mock-up for Adver.				
9/2	25	Printing/Binding - 1000 copies (50%)				
9/2	26	UPS - shipping/insurance to Printers	15.67			
9/7	27	500 Business Envelopes		7.95		
9/17	28	(3) Tapes, 100 Small Envelopes		10.10		
9/19	29	(2) Books - Aviation & Smuggling			8.90	
9/23	30	(6) Books - Aviation, Smuggling, Sailing			93.19	
9/23	31	Dep. (50%) on coupon typesetting				
9/23	32	Parking @ Brad. County Courthouse				
		Page Totals:	73.94	25.04	128.04	50.00

Figure 14-5

Expenses - Continued

Line	5 Book Production	6 Advertising	7 Food/Other	8 Travel/Lodging	9 Miscellaneous	10 Exhibit Fees	11 Dues/Fees	12
1							14.00	
2								
3		5.00						
4							45.00	
5								
6					7.61			
7	793.60							
8					7.59			
9					75.00			
10	20.67							
11		2.65						
12								
13								
14	719.20							
15	18.55							
16				66.80				
17			45.00					
18	702.25							
19								
20	12.72							
21								
22								
23					2.62			
24		5.57						
25	2370.00							
26								
27								
28								
29								
30								
31		21.20						
32					1.00			
	4636.99	32.42	45.00	66.80	93.82	0.00	59.00	0.00

Figure 14-6

SPECIAL FOURTH CLASS RATE
CONTENTS: BOOKS-MANUSCRIPTS

RETURN POSTAGE GUARANTEED

Figure 14-7

and dozens of other items that are used on a day-to-day basis while running a business.

Resource Books: Includes annual subscriptions to writing and publishing magazines, such as *Writer's Digest*, *Publishers Weekly*, *Publishing Entrepreneur*, and *Small Press*, as well as writing and publishing books, such as *How To Write, Publish & Sell Your Own Aviation Books*, *The Self-Publishing Manual*, *Is There A Book Inside You?*, *1001 Ways To Market Your Books*, and *How To Promote & Market Your Own Books*.

Freelance Help: Includes the hiring of outside help to do various freelance jobs, such as editing, artwork, photography and cover design.

Book Production: Includes all costs associated with typesetting, paste-up, printing and binding.

Advertising: Includes all costs associated with running advertisements in magazines and newspapers, producing flyers and brochures, and producing advertising displays for use at trade shows and special events.

Food/Other: Includes food, tips and other similar expenses incurred while doing business.

Travel/Lodging: Includes, motel and hotel expenses, travel expenses, road tolls, parking fees and other similar expenses incurred while doing business.

Miscellaneous: Includes items that either don't come up on a regular basis or aren't covered by any of the other categories, such as annual bank vault rental fees, monthly bank service charges and annual post office box rental charges.

Exhibit Fees: Includes booth rental fees and registration fees for book shows and trade shows.

Dues/Fees: Includes membership dues and fees for publishing trade associations, such as the National Association of Independent Publishers, as well as membership dues and fees for writing organizations, such as the National Writers Association.

The second ledger in my original record-keeping system was the sales ledger, which kept track of all book sales (including taxable sales and non-taxable sales). Figure 14-8 shows a page from this ledger. The nine column headings that I used in my original system were: Date, Purchaser, Place Of Purchase, Discount, Check/Cash, Taxable, Non-Taxable, Florida 6% Tax, Shipping. The following paragraphs are brief descriptions of each heading listed in my original sales ledger.

Date: Records the day, month and year of the sale.

Purchaser: Records the name of the purchaser. This can be either a person, retail business, book wholesaler, book distributor, library or a special event/organization. For example, looking at Figure 14-8 you can see that on February 11, 1992, I sold numerous books to an aviation retailer that was selling my books to pilots. A couple of weeks later, on February 24, 1992, I sold two books to a book wholesaler that sells my books to libraries. As indicated, the books that were sold to the wholesaler were eventually returned to me

Additional Income: $50 - 4/10/92
Sun 'n Fun Guest Speaker

Date	Purchaser/Place of Sale/Discount	C/$	Taxable	Non-taxable	FLA 6% tax	Shipping	
2/11/92	AMR Combs FLL - FL - 40%	C		11.37			1
2/11/92	AMR Combs FLL - FL - 40%	C		11.37			2
2/11/92	AMR Combs FLL - FL - 40%	C		11.37			3
2/11/92	AMR Combs FLL - FL - 40%	C		11.37			4
2/11/92	AMR Combs FLL - FL - 40%	C		11.37			5
2/11/92	AMR Combs FLL - FL - 40%	C		11.37			6
2/11/92	AMR Combs FLL - FL - 40%	C		11.37			7
2/11/92	AMR Combs FLL - FL - 40%	C		11.37			8
2/11/92	AMR Combs FLL - FL - 40%	C		11.37			9
2/22/92	Josh & Debbie Goldwitz - NY	$		18.95		1.50	10
~~2/24/92~~	~~Baker & Taylor Southern - GA - 50%~~	C	Returned	~~9.43~~		~~1.00~~	11
~~2/24/92~~	~~Baker & Taylor Southern - GA - 50%~~	C	Returned	~~9.44~~		~~1.00~~	12
3/3/92	Andrea - FL	$	18.95		1.13		13
3/5/92	Linda - FL	C	18.95		1.13		14
3/17/92	John Steadman - FL	$	18.95		1.13		15
3/25/92	William Murphy - AM - NY	C		17.95			16
4/9/92	Dave Prather - PBCC - FL	$	18.95		1.13		17
4/9/92	Carolyn Martin - PBCC - FL	C	18.95		1.13		18
4/10/92	Sun 'n Fun Fly-In - FL	C	18.95		1.13		19
4/10/92	Sun 'n Fun Fly-In - FL	$	18.95		1.13		20
4/10/92	Sun 'n Fun Fly-In - FL	$	18.95		1.13		21
4/10/92	Sun 'n Fun Fly-In - FL	$	18.95		1.13		22
4/10/92	Sun 'n Fun Fly-In - FL	$	18.95		1.13		23
4/10/92	Sun 'n Fun Fly-In - FL	$	18.95		1.13		24
4/10/92	Sun 'n Fun Fly-In - FL	$	18.95		1.13		25
4/10/92	Sun 'n Fun Fly-In - FL	$	18.95		1.13		26
4/10/92	Sun 'n Fun Fly-In - FL	$	18.95		1.13		27
4/10/92	Sun 'n Fun Fly-In - FL	$	18.95		1.13		28
4/10/92	Sun 'n Fun Fly-In - FL	$	18.95		1.13		29
4/10/92	Sun 'n Fun Fly-In - FL	$	18.95		1.13		30
4/15/92	Shelly Sherman - AM - IL	C		17.95		1.50	31
4/22/92	Linda Dickerson - IL	N/A	Complimentary	—	See Expenses		32

Figure 14-8

for a refund. The last entry shown in Figure 14-8, which took place on April 22, 1992, was not a book sale. Rather, it was for a complimentary gift copy of *Black Falcon* that I sent to the woman who had arranged my first aviation-related speaking engagement at the annual Sun 'n Fun EAA Fly-In at Lakeland-Linder Regional Airport. Now take a look at the notation at the top of Figure 14-8 that reads *Additional Income: $50 - 4/10/92 Sun 'n Fun Guest Speaker*. As a guest speaker for the After Dark Evening Program at the Sun 'n Fun EAA Fly-In, a $50 check was sent to me to cover travel costs.

I would like to point out that Figure 14-8 lists the sales that took place during a two-month period that began on February 11, 1992 and ended on April 22, 1992. In fact, Figure 14-8 doesn't even show all of the sales that took place during that time because the aviation retailer had purchased ten copies of *Black Falcon*, meaning that one of their entries was on the previous page of the sales ledger. My point here is that the ledger page shown in Figure 14-8 is just a randomly chosen example from hundreds of pages. Some months are good and some months are bad as far as sales are concerned. This is the way it is with *all* businesses - not just self-publishing ventures. During the two-month period shown in Figure 14-8, I did $540.70 in sales on a part-time basis selling one product (*Black Falcon*). At the time I wasn't even promoting and marketing my novel very much because I was too busy writing my second book (*How To Get Published - Guaranteed*). Throughout my years as a self-publisher, I've had periods where I've gone an entire month without a single book sale, but I've also experienced periods when I've brought in more than $1,000 in a single week. On one occasion I brought in $2,700 during a week-long trade show and on another occasion I sold $900 worth of books after giving a two-hour talk (which is my current one-day sales record). Not bad for an aircraft refueler who earns roughly $500 per week at his full-time job. Now that I have more than one product to sell, plus the experience to actively promote and market my entire product line, the possibilities are endless. And remember, I'm still doing this on a part-time basis. Things can only get better once I start self-publishing on a full-time basis.

Place Of Purchase: Records the state in which the sale took place, indicating (for retail sales only) whether or not Florida sales tax was to be collected for the sale. For instance, the state codes listed in Figure 14-8 show that sales took place in Florida, New York, Georgia and Illinois (the state codes being FL, NY, GA and IL) during this particular two-month period. Florida sales tax was only collected for the retail sales that took place in Florida.

Discount: Records the amount of discount extended to commercial customers. In Figure 14-8, for example, a 40% discount was given to the aviation retailer and a 50% discount was given to the book wholesaler. Notice also the two sales that were made on March 25, 1992 and April 15, 1992. Although the retail price of *Black Falcon* was $18.95 at that time, those two sales were made at $17.95, yet the books were not sold at a discount. What is the reason for the discrepancy of those two mail-order sales (one from New York and the other from Illinois)? Well, the Summer 1989 issue of *Airliners* magazine had an excellent book review about *Black Falcon*. I had also run an advertisement in that issue (plus the next three quarterly issues). Back then, the retail price of *Black Falcon* was $17.95, therefore that was the price that was listed in my advertisement. Obviously, the two people who sent me those mail-orders had read my advertisement in back issues of *Airliners* magazine (which were still being sold at some aviation hobby shops around the country) and sent me checks for $19.45 (which included $17.95 for the book, plus $1.50 for shipping). They had no way of knowing that the price had increased to

$18.95 and I wasn't going to risk losing the two sales by sending them a letter asking for an extra dollar, so I shipped the two books. How did I know that both orders were a result of the advertisement in *Airliners* magazine? Because the address in my advertisement had been coded as: Tri-Pacer Press, P.O. Box 840111, Dept. AM, Pembroke Pines, Florida, 33084. The *Dept. AM* had nothing to do with my address at all, but rather it was a coded tag that let me know where the orders were coming from - in this case, from my ad in *Airliners* magazine. Every promotion and marketing deal I do has its own coded tag. That way I know which ideas work and which ideas don't work. All I have to do is look at the envelopes that arrive in the mail, decipher the coded tag, and keep a running tab of the success or failure of each particular idea.

Check/Cash: Records the method of payment - either by check or cash. The letter *C* indicates a check, while the symbol *$* indicates cash. Once in a while, someone will pay with a money order, but this is rare. In those instances, I simply use the letters *M/O*. If a book was given away as a review copy or gift, I use the letters *N/A* for not applicable. The letter *X* that you see in this column on top of the other symbols indicates that thc orders were fulfilled (meaning that the books were shipped out).

Taxable: Records the price of all taxable retail sales made within Florida.

Non-Taxable: Records the price of all non-taxable retail sales made outside of Florida, plus all tax-exempt commercial sales made in all states, including Florida.

Florida 6% Tax: Records the amount of tax collected from all retail sales made within Florida.

Shipping: Records the amount of money collected for shipping charges.

The third ledger in my original record-keeping system was the vehicle mileage ledger, which kept track of the mileage I put on my truck while doing company-related business. Why? Because all business-related mileage is tax deductible. However, in order to legally be allowed to do this, accurate records must be kept of a vehicle's starting and ending odometer readings each time that vehicle is used for business-related purposes. Figure 14-9 shows one of the pages from my original vehicle mileage ledger.

Keeping track of mileage is easy. Before starting out on a company-related drive, just write down the date, the reason for the drive and the starting odometer reading. After the drive, write down the ending odometer reading, subtract the beginning reading from the ending reading, then write down the total mileage driven for that particular trip.

The mileage adds up fast. For instance, look at the entry dated February 6, 1992 in Figure 14-9. I drove to Bristow Graphics, then back home, for a round trip of fifteen miles. Now look at the entry dated April 10, 1992 to April 11, 1992, at which time I drove to Central Florida to give a talk to pilots at the Sun 'n Fun EAA Fly-In. Total mileage for the round-trip was 462 miles. Notice on this entry that the odometer readings were out of sequence from the rest of the entries. This was because my truck was experiencing major mechanical problems, so I drove my wife's car instead. You can see in Figure 14-9 that I wrote the notation *Honda* next to the total mileage figure. Now look at the entry dated August 7, 1992. Again, the odometer readings were out of sequence from the rest of the entries. Why? Well, due to continuing problems with my truck (I'm talking *major* problems with a *non-Chevy* truck), I was forced to purchase a new vehicle (this time I went back to a Chevy) so that I'd have reliable transportation all the time. The notation *New Truck* can be seen next to the total mileage figure for that entry.

Recording mileage might seem like a hassle to you, but without these detailed records the Internal Revenue Service

(11) Mixed 1991/1992 Mileage

1991/92		Reason	Starting	Ending	Total
11/7	1	Hollywood Fed, Pearl's Art, Office Depot	67591	67636	45
11/15-11/17		Miami Book Fair Int'l.	67723	67850	127
11/18	3	Hollywd. Fed., UPS	67850	67863	13
11/20	4	Palm Beach Illus., Libertes Books, Hlwd Fed.	67947	68070	123
12/6	5	Vault	68370	68385	15
1/6	6	Hollywood Fed.	69138	69148	10
1/7	7	Books & Books, Neway Printing	69180	69238	58
1/9	8	Neway Printing	69290	69298	8
1/13	9	Neway Printing	69507	69515	8
1/13	10	Neway Printing	69542	69550	8
1/14	11	Neway Printing	69656	69664	8
1/27	12	Hollywood Fed.	69990	70000	10
1/28	13	Southeast Graphics	70027	70042	15
1/30	14	BookFest Kickoff, Bristow Graphics, Vault	70081	70199	118
2/5	15	Bristow Graphics	70375	70390	15
2/6	16	Bristow Graphics	70437	70452	15
2/10	17	Office Depot, BCC Library	70541	70547	6
2/11	18	AMR Combs	70558	70576	18
2/19	19	Book Group Meeting	70737	70765	28
2/25	20	Hollywood Fed	70949	70959	10
4/9	21	PBCC-National Book Week	72120	72218	98
4/10-4/11		Sun 'n Fun Fly-In, Lakeland, FL	90061	90523	462 Honda
4/15	23	Hollywood Fed.	72335	72341	6
4/20	24	Vault, Hollywood Fed, Office Depot	72407	72425	18
4/24-4/26		Book Fest	72537	72863	326
4/27	28	Hollywood Fed.	72863	72873	10
5/6	27	Vault	73223	73238	15
5/11	28	Hollywood Fed	73353	73363	10
5/20	29	Vault	73661	73676	15
7/8	30	Vault	74624	74639	15
8/7	31	Office Depot	319	324	5 New Truck
9/9	32	AMR Combs	2014	2032	18
		Page Totals:			323/1333

Figure 14-9

will not allow you to take the tax deductions that are legally yours to take. And let me tell you, these deductions add up fast. When I started Tri-Pacer Press in 1987, the IRS was allowing a deduction of $0.22 per mile. By 1992 it was up to $0.28 per mile. By 1996 it was up to $0.31 per mile. And it keeps going up all the time. During my early years as a self-publisher, I was averaging about 1,500 to 2,000 miles per year of business-related driving, but that figure has now increased drastically. In 1996 alone I drove 3,598 miles while doing company-related business. By multiplying 3,598 miles by $0.31 per mile, I received a $1,115 tax deduction on my 1996 income taxes. So you can see what a sizable tax savings business owners can get if they keep proper records.

The final portion of my original record-keeping system involved four file folders that contained important documents relating to my business. (Actually, I had many more than just four file folders in my filing system, but the others had nothing to do with financial record-keeping.) The headings on these four file folders were: Bank Statements, Florida Sales Tax Reports, Purchase Orders, Receipts Current Year. The following paragraphs are brief descriptions of the contents of each file folder.

Bank Statements: Every month Tri-Pacer Press receives a statement from the bank that lists a description of each transaction made during the previous month, plus a check reconciliation summary and an account summary. After I balance this statement against my own records, I file it away in chronological order in the Bank Statements file folder. Figure 14-10 shows a copy of a monthly bank statement from my business account. Notice that the name on the statement is listed as *Richard Cook D/B/A Tri-Pacer Press*. This stands for Richard Cook doing business as Tri-Pacer Press, which is the legal description of my business.

Florida Sales Tax Report: Every three months (quarterly) Tri-Pacer Press must send in a Sales And Use Tax Re-

HOLLYWOOD FEDERAL

000270

RICHARD COOK D/B/A
TRI-PACER PRESS

DEMAND ACCOUNT - 47
PERIOD BEGINNING 01/01/89
PERIOD ENDING 01/31/89
PAGE NUMBER 1

DATE	DESCRIPTION	AMOUNT	BALANCE
01/01	BEGINNING BALANCE		1,207.79
01/01	SERVICE CHARGE FOR 12/01 THRU 12/31	4.95	1,202.84
01/03	DEP AT EMERALD HILLS	151.63	1,354.47
01/09	DEP AT EMERALD HILLS	289.74	1,644.21
01/11	CHECK: 1009	101.68	1,542.53
01/17	DEP AT EMERALD HILLS	885.78	2,428.31
01/17	CHECK: 1010	73.40	2,354.91
01/25	DEP AT EMERALD HILLS	72.23	2,427.14

************ C H E C K R E C O N C I L I A T I O N S U M M A R Y ***********

CHECK	DATE	REFERENCE NUMBER	AMOUNT
1009	01/11	07900173	101.68-
1010	01/17	08200060	73.40-

************************* A C C O U N T S U M M A R Y *************************

BEGINNING BALANCE	1,207.79
PLUS 4 CREDITS	1,399.38
	2,607.17
LESS 2 DEBITS	175.08-
	2,432.09
LESS ACCOUNT CHARGES	4.95-
ENDING BALANCE	2,427.14

AVERAGE COLLECTED BALANCE FOR JANUARY $1,859.54

Figure 14-10

turn to the Florida Department of Revenue, along with a check for the amount of sales tax I collected during the previous three months. Figure 14-11 shows the quarterly Sales And Use Tax Return from the period that began on October 1, 1988 (when *Black Falcon* was first published) and ended on December 31, 1988 (the end of the first quarter that *Black Falcon* was available to the public). During that three-month period, I sold $3,373.72 worth of books. Of that amount, $1,489.52 was taxable at a rate of 6% (which came to $89.37) and $1,884.20 was non-taxable. Also, I had purchased $257.67 worth of taxable supplies (shipping envelopes, shipping tape and other related items) that should have been taxed by the supplier when I purchased the items, but weren't. Because of this, I still owed $15.46 in user taxes for the $257.67 worth of supplies. This brought the total amount of taxes owed up to $104.83, minus a refund of $3.15 (called a collection allowance, which the Florida Department of Revenue offers to businesses that collect taxes), which brought the actual amount owed to the

STATE OF FLORIDA DEPARTMENT OF REVENUE SALES AND USE TAX RETURN DR-15

	1. GROSS SALES	2. EXEMPT SALES	3. TAXABLE AMOUNT	4. TAX COLLECTED
A SALES	$3373.72	$1884.20	$1489.52	$89.37
B TAXABLE PURCHASES			$257.67	$15.46
C AGRICULTURAL EQUIP.				
D TRANSIENT RENTALS				

TRANSIENT RENTAL RATE SURTAX RATE

CERTIFICATE NO. SIC FEI/SSN PERIOD

TRI-PACER PRES COPY

COOK RICHARD

PEMBROKE PINES FL

5	TOTAL AMOUNT OF TAX COLL.	$104.83
6	LESS LAWFUL DEDS.	
7	TOTAL TAX DUE	$104.83
8	LESS EST. TAX PAID	
9	PLUS EST. TAX DUE CURRENT MO.	
10	AMOUNT DUE	$104.83
11	LESS: COLL. ALLOW.	3.15
12	PLUS: PENALTY ($5.00 MINIMUM)	
13	PLUS INTEREST	
14	AMOUNT DUE WITH RETURN	$101.68

RETURN IS DUE BE SURE TO SIGN AND DATE THE REVERSE SIDE.

DO NOT MAIL THIS COPY

Check #1009

THIS IS YOUR COPY OF THE TAX RETURN - RETAIN FOR YOUR RECORDS

Figure 14-11

state down to $101.68 - an amount that I sent to the Florida Department of Revenue on January 4, 1989, along with the original copy of the Sales And Use Tax Return. The duplicate copy of this form was then filed away in the Florida Sales Tax Report file folder. If you now look back at the bank statement shown in Figure 14-10, you will see that the check I sent to the Florida Department of Revenue (check #1009) was cleared by their bank and withdrawn from my bank on January 11, 1989.

Purchase Orders: When commercial customers (such as retailers, wholesalers and distributors) place their book orders, they usually come in the form of purchase orders. Purchase orders vary in their degree of sophistication, depending on the business that placed the order. They can either be hand-written or computerized. Although each company uses its own format, they all provide the same basic information. Figure 14-12 shows a purchase order I received back in mid-1989 for one copy of *Black Falcon* from a bookstore located in Michigan. One of their customers probably read a book review about my novel and decided to order the book. Notice that the purchase order lists the date, the method of shipment (UPS), the quantity ordered, the ISBN Number, the type of book (cloth, which means hardcover), the title, the author and the retail price. This bookstore did not fill in the discount information, so I added it to the purchase order. A discount of 25% was applied since they only purchased one copy for resale, plus I added in $2.50 for the UPS shipping charges. This store had sent a blank check, which I filled out for the total amount owed to me, which came to $15.97. As I already stated, every purchase order looks different, but they all include the same information. After shipping out each order, I file purchase orders in the Purchase Order file folder.

Receipts Current Year: Whenever an expense is incurred while doing company-related business, I always get a

Ship & Bill To:

KENTWOOD, MICHIGAN 49508

PURCHASE ORDER
X 11229 V
DATE
7/28/89
SHIP VIA
UPS

Attn: SPECIAL ORDER DEPARTMENT

TO: Tri Pacer Press
P.O. Box 840111
Pembroke Pines, FL 33084

P = Paper
C = Cloth

QTY	ISBN	TYPE	TITLE	AUTHOR	PRICE
1	0-961999152	c	Black falcon	Alden, C.	17.95
				Discount -25%	<4.48>
				Sub-total	13.47
				Freight UPS	2.50
				Total →	15.97

- [ ] **BILL TO OUR ACCOUNT:** ____________
- [x] **PREPAID BLANK CHECK ENCLOSED**
- [x] **CHECK #** 13522 **ENCLOSED FOR $** ____________
- [ ] **BACKORDER IF NECESSARY**
- [x] **DO NOT BACKORDER**

Please direct all correspondence concerning this order to SPECIAL ORDER DEPARTMENT

Figure 14-12

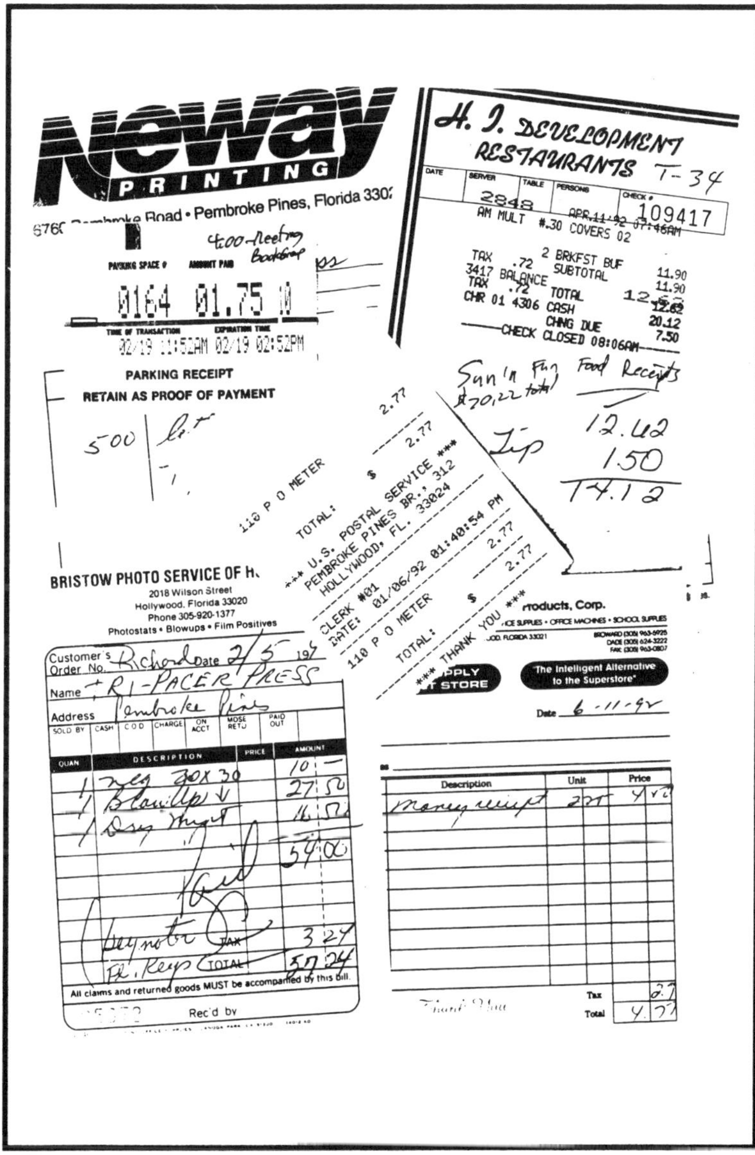
Neway PRINTING
Pembroke Pines, Florida
H. J. DEVELOPMENT RESTAURANTS
PARKING RECEIPT
RETAIN AS PROOF OF PAYMENT
U.S. POSTAL SERVICE
PEMBROKE PINES BR., 312
HOLLYWOOD, FL. 33024
THANK YOU
BRISTOW PHOTO SERVICE
2018 Wilson Street
Hollywood, Florida 33020
Phone 305-920-1377
Photostats • Blowups • Film Positives
The Intelligent Alternative to the Superstore
All claims and returned goods MUST be accompanied by this bill.

Figure 14-13

receipt - even for something as trivial as a $1.25 turnpike toll. Why? Because I need them for end-of-the-year tax purposes. Without receipts, the Internal Revenue Service will not allow a business to deduct their business expenses. Sure, it's a hassle to always ask for a receipt, but it all pays off when tax season rolls around. Figure 14-13 shows a montage of various receipts I collected during 1992. After an expense is incurred, I enter it into the expense ledger, then file away the receipt in the Receipts Current Year file folder.

Good record-keeping is important for many reasons, but probably the two most important ones are for annual income tax purposes and knowing the health and wealth of a business. Without a good record of your expenses, sales and mileage (and the paperwork to back it all up), it would be impossible to accomplish either of the above necessities. Because I am not a tax expert, I suggest that you find a reputable accountant to help you file your annual income tax forms once you start your own publishing company. Obviously, this person will want to see all of your records, so be sure to begin your record-keeping right from the start.

As you can see from this chapter, keeping good records is not very difficult. In fact, if you can write a book, you can keep records - it's that simple!

CHAPTER FIFTEEN

PROMOTING & MARKETING YOUR BOOK

This chapter is about promoting and marketing your book (what I consider to be the fun part of self-publishing). Promoting and marketing involves various things, such as press releases, book reviews, personal interviews, speaking engagements, sales trips, book fairs, special events (both literary and non-literary) and much more. All of this fun stuff is geared towards one goal - selling copies of your book. How is this accomplished? Through ingenuity, imagination and a determination to succeed. I have done it with *Black Falcon* just as thousands of other authors have done it with their books.

Most self-published books will not become bestsellers, but sometimes they do. Examples include *The Tale Of Peter Rabbit*, *Bartlett's Quotations*, *Lady Chatterly's Lover*, *Ulysses*, *The First Whole Earth Catalog*, *The One-Minute Manager*, *Tarzan And The Apes*, *Winning By Intimidation* and *The Adventures Of Huckleberry Finn*.[53] The rights to most of these books were eventually picked up by major publishers, but each one of them came out originally as a self-published book and each one of them achieved its initial success through a promotion and marketing campaign set up by its respective author.

One of the key ingredients to promoting and marketing a book involves *selling yourself as an author*. This doesn't mean paying a publicity agent thousands of dollars to publicize your book, because most self-publishers don't have this kind of money. What it *does* mean is that you, as the author, must spend time and energy seeking out all the free publicity that's available to every author - and let me tell you, there's a lot of free publicity available to those who are willing to put forth the effort to get it. Keep in mind, though, that all of this publicity cannot be acquired overnight, so don't get yourself stressed out trying to do everything all at once. Pace yourself. Don't burn yourself out by going beyond your personal limits. But *do* be continual and steady in your search for publicity. I've been searching for and acquiring free publicity since 1988.

How is this free publicity acquired? Primarily by sending out press releases and review copies of your book to magazines and newspapers, but also by seeking out people who are interested in the same subject as your book, which is accomplished by attending conventions, trade shows, association meetings and special events that deal with your book's subject matter.

To highlight this process, let me use *Black Falcon* as an example. When my novel was originally published back in 1988, I mailed press releases and review copies to dozens of aviation and sailing magazines and to several Florida newspapers. By doing so, I was targeting the three most likely markets for my novel (pilots, sailors and Florida residents), hoping that at least a few of the magazines and newspapers would print book reviews about *Black Falcon*. As it turned out, I ended up getting a lot of media attention.

I also spent a lot of time hanging out at airports, marinas, pilot shops, boating supply stores, air shows and boat shows. Much of my time at these places was spent talking to people about flying and sailing. Believe me, there's not much I enjoy doing more than talking about flying and sailing, except actually doing the real thing. The fact that I was an air-sea adventure novelist would sometimes arise during my conversations with pilots and sailors, and a good many of these people would then ask how they could purchase an autographed copy of my book. I would give them the name and address of Tri-Pacer Press and, low and behold, orders would then appear in the mail - at full retail price! These people, in turn, would tell other people about my book, bringing in more orders. Word-of-mouth is a great sales tool. Use it to your advantage during your search for publicity.

I had mentioned a process known as networking back in Chapter Eight - how one marketing idea leads to other marketing ideas, which then leads to new personal contacts, which leads to additional marketing ideas, and so on and so forth. Well, Chapter Fifteen will tell you how to put this networking process into motion. Once you start, however, don't expect every marketing idea to work out, and don't expect the ones that *do* work to all fall into place at the same time. I've had plenty of failures on my path to success. In fact, it took me six years to finally expand my publishing business into something that will someday lead me to financial independence (*I hope*). The important point here is that I had the willingness to try new ideas without being afraid to fail once in a while. You, too, must be willing to promote and market your book without being afraid to fail - otherwise you will fail before you even begin. I know a few self-publishers who couldn't sell very many copies of their books because they took the attitude that since they were the authors, they didn't have to promote and market their books. Well, that kind of attitude will cause any business to fail, whether it be an independent publishing company, a kitplane manufacturer, a fixed base operation, or whatever. You, as a self-publisher, are the owner of a business. Your product is your book. You must be the one who promotes and markets your product.

Indeed, promotion and marketing are both very necessary in order to sell books. As a matter of fact, this is where self-publishers have a clear advantage over major publishers (unless you are a mega-author or a celebrity of some sort). Why? Because major publishers spend most of their time and money on the promotion and marketing of their few bestsellers (or what they hope will become bestsellers), which leaves very little time and money left over for the majority of their books. This isn't a new policy. It's been this way for decades. A very good example is William Faulkner. Although he was eventually awarded one Nobel Prize and two Pulitzers for his later writings, his first four novels sold an average of only 2,000 copies each.[54] He just wasn't well-known enough during the early part of his career for the publishers to push his books. The sales of *my* first two books, on the other hand, will each go far beyond 2,000 copies apiece. (In fact, *Black Falcon* has already gone past this mark). The same will hold true for all of my books. How do I know this? Because I have the willingness to promote and market my books continuously and will never take them out of print.

The rest of this chapter will be an overview of how I sold out the first printing of *Black Falcon* during my first two years as a self-publisher, plus an assortment of additional promotion and marketing efforts that have taken place since selling out the first printing. These are not hypothetical examples (as you would find in a traditional textbook). These cases studies are the real thing! As you read through all of my personal experiences, keep the following thoughts in mind. First of all, promotion and marketing should be a continual effort. This doesn't mean that you have to work full-time on it (I surely don't), but you should always have something on the burner. I'm still selling copies of *Black Falcon* because of my continual efforts. Secondly, free publicity should always be sought out before resorting to paid advertising. Not only is advertising very expensive, but the general public believes book reviews more than they believe advertising copy. I'd estimate that 95% of the publicity cast upon *Black Falcon* was due to free publicity, as compared to only about 5% being from advertisements that I had placed in various publications. And thirdly, there are so many hundreds (even thousands) of ways to promote and market books, that I couldn't possible squeeze everything I've tried into just this one chapter of *How To Write, Publish & Sell Your Own Aviation Books* (which is why I'm working on a sequel that will go into more detail regarding this area of interest, titled *How To Promote & Market Your Own Books*).

How did I learn about all of these ideas for promoting and marketing books? Well, many of them came from a book titled *1001 Ways To Market Your Books* by John Kremer. (To order a copy of *1001 Ways To Market Your Books*, please refer to the *Writer's Book Mart Catalog* in the appendix section of *How To Write, Publish & Sell Your Own Aviation Books*.) Other ideas came from my imagination. Some even came from various self-publishers who had shared their successes with me. Figures 15-1 and 15-2 show a list of some of the ways I promoted and marketed *Black Falcon* during my first five years as a self-publisher.

Now let's get on with this chapter and discuss how to get free publicity, which primarily comes from book reviews and news stories in magazines and newspapers. How does a self-publisher get this kind of free publicity? By sending out press releases and review copies in what is known as a press kit. The next several pages will show you in detail how I put together a workable press kit for *Black Falcon*. It's not hard to do, but it does take time to research each of the publications that will be receiving the press kits.

I began researching magazines and newspapers during the final production stage of *Black Falcon* by spending a day at a local library flipping through *The Standard Periodical Directory* (which I told you about in Chapter Eight). I

Promotion & Marketing activities for Black Falcon (October 1988 - April 1994):

(Note: These listings are not in chronological order.)

Airliners International Memoribilia Trade Show, Booksigning.

Mailing to Florida-based libraries.

Press Releases and Review Copies to various magazines and newspapers.

Little Professor Book Center, Booksigning.

Banyan Air Service Open House, Booksigning.

American Association of Airport Executives Aviation Symposium, Guest Speaker.

Aviation Day at Valkaria Airport, Booksigning.

Miami International Sailboat Show, Booksigning.

The Pilot Supershop Grand Opening, Booksigning.

Southern Florida Airline Historical Assoiation Memoribilia Trade Show, Booksigning.

Florida Publishers Association Convention And Trade Show, Booksigning.

Miami Book Fair International, Booksigning.

American Booksellers Association Convention via COSMEP participation.

American Library Association Convention via COSMEP participation.

Film Rights offered to actor John Travolta.

BookFest Of The Palm Beaches, Guest Panelist and Booksigning.

Mass-Market Paperback Rights offered to fifteen major New York publishing houses.

Mickler's Floridiana Book Catalog carried "Black Falcon" as a sales item.

Sailorman Marine Emporium, Booksigning.

Classified Ads in various magazines and newspapers.

Display Ads in various magazines.

National Book Week at BCC Library, Guest Speaker and Booksigning.

National Library Week at PBCC Library, Guest Speaker and Booksigning.

Palm Beach Literary Forum & Author's Luncheon, Honored Guest and Booksigning.

The Backroom Writers Group, Guest Speaker and Booksigning.

Figure 15-1

Promotion & Marketing activities for Black Falcon (October 1988 - April 1994):

Sun 'n Fun EAA Fly-In, Guest Speaker, Forum Instructor and Booksigning.

Numerous sales of "Black Falcon" via the following aviation supply stores, marine supply stores and independent bookstores:

The Flight Department	Liberties Fine Books & Music
Gifts Of Flight	AMR Combs
Borders Book Shop	Jolie's Books
Airpark Aviation	Sailorman Marine Emporium
Little Professor Book Center	Ballen Booksellers International
Miami Aviation	Page Avjet
Scholarly Book Center	Duval Books & Cards
Yankee Book Peddler	Savco Flying
The Armchair Sailor	Pier 66 Marine Service Center
Schuler Books	Hawley-Cooke Bookseller
Island City Flying Service	Mariner's Cove Marine Supply
Food For Thought Book Store	Book Fair
Gateway Bookstore	The Pilot Supershop
Cody's Books	The Book Nook
Plane Things	Banyan Air Service
Jet Center	Downtown Book Center
Holiday Flying Club	Sidney Johnson Bookseller
Aero Gem	Avico
Boca Aviation	City News & Books
Annie's Book Stop	Professional Flight Crew Services
FIT Aviation	Hangar One Bookstore
Aero-Supply	Bluewater Books & Charts

Figure 15-2

wrote down the names and addresses of aviation and sailing magazines, as well as Florida-based newspapers. I also got the names of specific editors and staff writers for each of these publications.

After completing my research, I put together a group of approximately eighty publications that would be receiving press kits. Each of the press kits would contain an individualized cover letter, a press release, an author fact sheet, a dustjacket from *Black Falcon*, and a self-addressed stamped response card. I then divided the eighty publications into two separate groups - those that would receive a complimentary review copy of *Black Falcon* in their press kits and those that would not get a review copy of the book. The reason I did this was because I could only afford to use 5% of the first printing of *Black Falcon* as free copies in order not to lose too much sales revenue by taking these books out of my sales inventory. That 5% came to fifty copies.

After deciding which publications were most likely to print book reviews as opposed to news stories, I selected the fifty publications that would get a copy of *Black Falcon*. Then it was only a matter of putting together the press kits, mailing them out and waiting for the responses to come in. To make a long story short, dozens of publications responded to my press kits - some saying they would definitely do a write-up about *Black Falcon*, others saying they would not do a write-up about *Black Falcon*, and still others saying a write-up was possible (but not guaranteed). Also, a few of the publications that had not received a review copy asked to have one sent to them, which I naturally obliged. One of those publications ended up writing the second best book review that was ever written about *Black Falcon*.

Before going into detail about the responses I received regarding my press kits, let's first discuss the actual contents of the press kits. Probably the most important aspect of any press kit is the cover letter. This letter should be addressed to a specific editor or staff writer. It should also be short (no more than one page) and should be worded specifically towards the readership of each individual publication. In other words, more than one version of the letter might have to be written. Figures 15-3, 15-4 and 15-5 show the cover letters I used for the three different types of publications I sent press kits to (which were aviation-oriented magazines, nautical-oriented magazines and Florida-oriented newspapers). You can see in these actual examples that each letter was addressed to a specific editor or staff writer.

The first paragraph of each cover letter was the same because I used this portion of the letter to introduce my book. The first paragraph stated, "Tri-Pacer Press is proud to announce the publication of *Black Falcon*, a thrilling air-sea adventure novel by Chevy Alden. We are actively seeking book reviews, passage excerpts, noteworthy news mentions and author interviews. If [*publication's name*] is interested in printing any of the above items, please fill out the enclosed pre-stamped response card and drop it in the mail as soon as feasible."

The second paragraph of each cover letter was where I individualized the wording to reflect each publication's particular readership. This showed each editor and staff writer that I spent time doing research and that I didn't just pick their names out of a hat. In the end, the results were worth the effort.

The second paragraph in the letter I sent to *Airliners*, which is shown in Figure 15-3, stated, "*Black Falcon* and Chevy Alden should be of particular interest to your readers because of the novel's aviation-oriented subject matter and the author's background in the aviation industry. The book features old radial-engined airfreighters throughout the story line, as well as a few scenes with jet aircraft." The editorial content of *Airliners* dealt with the history of past and present airliners, especially old radial-engined transports like

TRI-PACER PRESS, P.O. Box 840111, Pembroke Pines, Florida 33084
Publisher of Quality Action-Adventure Novels

November 6, 1988

John Wegg, Editor
Airliners
P.O. Box 1439
Pollock Pines, CA 95726

Dear Mr. Wegg:

Tri-Pacer Press is proud to announce the publication of Black Falcon, a thrilling air-sea adventure novel by Chevy Alden. We are actively seeking book reviews, passage excerpts, noteworthy news mentions and author interviews. If Airliners is interested in printing any of the above items, please fill out the enclosed pre-stamped response card and drop it in the mail as soon as feasible.

Black Falcon and Chevy Alden should be of particular interest to your readers because of the novel's aviation-related subject matter and the author's background in the aviation industry. The book features old radial-engined airfreighters throughout the story line, as well as a few scenes with jet aircraft.

Thank you for considering our query.

Sincerely,

Richard Cook
Editor-in-Chief

ENC: Press Release, Author Fact Sheet, Complimentary Review Copy.

Figure 15-3

TRI-PACER PRESS, P.O. Box 840111, Pembroke Pines, Florida 33084
Publisher of Quality Action-Adventure Novels

October 26, 1988

John Ziegler, Editor
Waterfront News
1224 SW 1st Ave.
Ft. Lauderdale, FL 33315

Dear Mr. Ziegler:

Tri-Pacer Press is proud to announce the publication of Black Falcon, a thrilling air-sea adventure novel by Chevy Alden. We are actively seeking book reviews, passage excerpts, noteworthy news mentions and author interviews. If Waterfront News is interested in printing any of the above items, please fill out the enclosed pre-stamped response card and drop it in the mail as soon as feasible.

Black Falcon and Chevy Alden should be of particular interest to your readers because of the novel's South Florida boating-related subject matter (especially the chase scene on Ft. Lauderdale's Venice-like canal system) and the author's Broward County residency.

Thank you for considering our inquiry.

Sincerely,

Richard Cook
Editor-in-Chief

ENC: Press Release, Author Fact Sheet.

Figure 15-4

TRI-PACER PRESS, P.O. Box 840111, Pembroke Pines, Florida 33084
Publisher of Quality Action-Adventure Novels

November 7, 1988

David Hawkins, Staff Writer
Florida Keys Keynoter
3015 Overseas Highway
P.O. Box 158
Marathon, FL 33050-0158

Dear Mr. Hawkins:

Tri-Pacer Press is proud to announce the publication of Black Falcon, a thrilling air-sea adventure novel by Chevy Alden. We are actively seeking book reviews, passage excerpts, noteworthy news mentions and author interviews. If Florida Keys Keynoter is interested in printing any of the above items, please fill out the enclosed pre-stamped response card and drop it in the mail as soon as feasible.

Black Falcon and Chevy Alden should be of particular interest to your readers because of the novel's many scenes in the Keys and the author's love for the island chain. Chevy sails and relaxes in the Florida Keys quit often with his family.

Thank you for considering our inquiry.

Sincerely,

Richard Cook
Editor-in-Chief

ENC: Press Release, Author Fact Sheet, Complimentary Review Copy.

Figure 15-5

the ones in my novel. *Airliners* did an excellent book review about *Black Falcon*.

The second paragraph in the letter I sent to *Waterfront News*, which is shown in Figure 15-4, stated, "*Black Falcon* and Chevy Alden should be of particular interest to your readers because of the novel's South Florida boating-related subject matter (especially the chase scene on Ft. Lauderdale's Venice-like canal system) and the author's Broward County residency." The editorial content of *Waterfront News* dealt with Ft. Lauderdale boating issues. *Waterfront News* did an excellent book review about *Black Falcon*.

The second paragraph in the letter I sent to the *Florida Keys Keynoter*, which is shown in Figure 15-5, stated, "*Black Falcon* and Chevy Alden should be of particular interest to your readers because of the novel's many scenes in the Keys and the author's love for the island chain. Chevy sails and relaxes in the Florida Keys quit often with his family." The editorial content of the *Florida Keys Keynoter* dealt with topics about island life in the Keys. Notice the misspelling of the word *quite*. I left out the *e*. Despite this obvious blunder (caused by using an old electric typewriter that did not have a spell-checking program, followed by a lousy job of proof-reading on my part), the *Florida Keys Keynoter* did an excellent book review about *Black Falcon*.

With the above three examples illustrating how I individualized each cover letter, you should have a fairly clear idea of how to do the same thing for your book. It takes time and it takes research. It *does not*, however, take a typographical error to get the attention of an editor or staff writer, so don't repeat my mistake. (Now, of course, I use a personal computer that has a spell-checking program.)

I ended each of my cover letters by simply stating, "Thank you for considering our query." The *our* made them think that Tri-Pacer Press was more than just a one-person publishing company.

The second item in each of my press kits was a press release. A press release is nothing more than a standardized method of distributing information to the media - whether it be magazines, newspapers, radio or television. You should get used to the standardized format of a press release because you will be writing them throughout your entire publishing career whenever you have something newsworthy to publicize. Figure 15-6 shows the press release I enclosed with my *Black Falcon* press kits. Notice the simple format of this important sheet of paper - one page, double-spaced, wide margins, a contact person, the title of the book, the author's name, a three paragraph description of the book that tells the five *w's* (who, what, where, when and why), plus the price and ordering information. At the bottom of the press release is the symbol *-30-*, which indicates *The End.* Be sure to include this symbol at the bottom of your press release so that you will come across as being a professional. Compare the press release shown in Figure 15-6 with the one shown in Figure 15-7, which shows the press release I used for *How To Get Published - Guaranteed.* I used my computer to write and print this press release, so it has a much more professional look to it. In fact, this was the very first item I produced with my computer while learning how to typeset. Notice that I was able to justify (align) both margins and also adjust the type size of my headings. I am *so* glad that my wife forced my to finally get a personal computer after I finished writing *How To Get Published - Guaranteed.* Besides making my writing much easier, it has improved my promotion and marketing abilities as well.

Because every magazine and newspaper receives hundreds of press releases each day from all sorts of businesses, it is sometimes wise to make your press releases stand out by photocopying them on colored paper. For example, I photocopied the press release for *Black Falcon* on light blue paper. This seemed to work for me without any problem.

TRI-PACER PRESS, P.O. BOX 840111, Pembroke Pines, Florida 33084 • 305-432-9683
Publisher of Quality Action-Adventure Novels

OFFICIAL PRESS RELEASE

FOR IMMEDIATE RELEASE

Contact: Richard Cook

Title: BLACK FALCON

Author: Chevy Alden

Advanced Copies Available: October 31, 1988

Official Publication Date: December 15, 1988

BLACK FALCON is a 257 page air-sea adventure novel that authentically depicts how the drug trade destroys the human spirit within those involved. This violently shocking novel flashes back to a time when South Florida's nighttime sky was a shadowy place to be...the early 1980's...the boom years for aerial marijuana smuggling. Instant wealth was offered to any pilot willing to risk his life flying tramp air-freighters down to Colombia and back. But smuggling was a fool's game...a deadly game not to be reckoned with.

BLACK FALCON introduces Rob Jensen, a physically strong, but emotionally unstable doper pilot who decides to quit the drug trade at the wrong time. With nobody to turn to for help, he enters into a desperate struggle for survival against his distraught emotions, nature's elements and the deadly world of drug running.

This fast-paced thriller takes the reader from a sleazy Miami bar to a salty Key West tavern, from the cockpit of a drug plane to a clandestine airfield in the mountains of Colombia, from a tropical Bahamian cruise aboard a private sailboat to a climactic action-packed ending on Ft. Lauderdale's Venice-like canal system.

BLACK FALCON will be available in hardcover at select book stores or directly from the publisher for $17.95 (FLA. addresses add 6% Tax). Order ISBN 0-9619991-5-2.

See Author Fact Sheet for personal biography about Chevy Alden.

-30-

Figure 15-6

TRI-PACER PRESS, P.O. Box 840111, Pembroke Pines, Florida 33084-2111 • 305-432-9683

Publisher of Quality Titles Since 1988

NEWS RELEASE

For Immediate Release....

Contact: Richard Cook (305) 432-9683

Authors Can Now Guarantee The Publication Of Their Books

Hundreds of thousands of manuscripts get written each year by unpublished authors. Unfortunately, most of these authors will never experience the thrill of seeing their manuscripts turned into books. Chevy Alden faced this dilemma after completing his action-adventure novel ***Black Falcon***, which was rejected by 84 major publishing houses. Rather than giving up, Alden researched the publishing industry until he found the secret to guaranteed publication. Within a year of his discovery, Alden's tale of drug smuggling was successfully published in a hardcover edition, received dozens of favorable book reviews in magazines and newspapers, and even found its way onto the President's bookshelf in The White House.

With ***Black Falcon*** now coming out in its third printing, Alden has written a new book titled ***How To Get Published - Guaranteed*** (328 pages, $18.95). This comprehensive self-help manual explains in detail how authors can write, publish and sell their own books by starting independent publishing companies to compete right alongside the major publishers. Known as self-publishing, this cottage industry has seen rapid growth during the past few years with the advent of affordable desktop publishing software, personal computers and laser printers. By contracting out the actual book production process to specialized printers and binders, high-quality books can be manufactured for as little as $2 per copy for 300-page books that sell for $18.95 or more in stores.

There are other "how to" books about self-publishing available, but ***How To Get Published - Guaranteed*** explores this topic from a refreshingly new angle. Rather than being a dry textbook, Alden's manual takes a folksy "this is how I did it" approach to the subject and is enriched with more than 100 photos, figures, illustrations, publishing forms, documents and sample letters. Furthermore, this book describes the human side of the picture: the emotions, the joys, the frustrations, the successes, the failures, the do's and don'ts of putting a book into print for the first time. Alden takes readers on an autobiographical journey through his first five years as a novice self-publisher, while at the same time giving readers the encouragement they may need to self-publish their own books.

Alden is not alone in his endeavors. Thousands of authors are creating prosperous writing careers for themselves via self-publishing. Using captivating names such as Prospector Press, Black Hawk Publishing and Killer Whale Productions, these enterprising authors are busily arranging speaking engagements, autograph sessions and personal appearances at book shows and other literary events. Alden, who lectures often, sums up self-publishing by saying that, "In today's world of publishing superconglomerates, you have to take your writing career into your own hands if you want to succeed. Besides, it sure beats waiting around for rejection letters."

Available at bookstores or directly from Tri-Pacer Press, POB 840111, Pembroke Pines, FL 33084-2111.

-30-

Figure 15-7

The third item in each of my press kits was an author fact sheet. An author fact sheet is nothing more than a summary of an author's writing career. When you first start out as a writer, your author fact sheet might be very short, so fill up the empty space with your interests, hobbies and background information. After your writing career takes off, your author fact sheet will expand with each new accomplishment, at which point you can remove your interests and hobbies. To see how to write an author fact sheet, turn back to Figures 1-1 and 1-2 in Chapter One. Those two pages show the author fact sheet that I used back in 1994 when *How To Get Published - Guaranteed* was just about finished. Keep in mind that all of the accomplishments listed in Figures 1-1 and 1-2 were achieved *without* the help of a publicity agent. Also keep in mind that *only* my major achievements were included on those two sheets of paper. If I had listed every minor achievement, my author fact sheet would have been twice as lengthy. As it turned out, I eventually realized that even a two-page author fact sheet was too long, so I condensed it down to a single sheet of paper. Figure 15-8 shows my author fact sheet from 1998, the year *How To Write, Publish & Sell Your Own Aviation Books* was published. Notice that instead of calling it an author fact sheet, I fancied up the heading by calling it an author biography. Either heading is fine because they both mean the same thing.

The fourth item in each of my press kits was a dustjacket from *Black Falcon*. That was the main reason I had extra dustjackets printed up during the production phase of publishing my novel. By including a book cover with each of your press kits, you will be letting the editors and staff writers see what your book actually looks like, which is especially important for the publications that won't be getting a review copy of the book.

The fifth item in each of my press kits was a self-ad-

AUTHOR BIOGRAPHY

CHEVY ALDEN lives in Broward County, Florida | DOB: May 22, 1958

BACKGROUND	Aircraft Refueler, Substitute Teacher, Pilot, Sailor, Adventurer, Thrill-Seeker.
PUBLISHED WORKS	*Black Falcon*, published by Tri-Pacer Press (1988), second printing (1991), third printing (1996). An air-sea novel about drug smuggling in South Florida. This 257-page thriller was chosen as "Book of the Month" by *US Aviator* and received excellent reviews from *Airliners*, *Biscayne Bay Times*, *Cruising World*, *Florida Keys Keynoter*, *Miami Herald*, *Private Pilot* and *Waterfront News*. *How To Get Published - Guaranteed*, published by Tri-Pacer Press (1994). An easy-to-understand text that describes book publishing from an insider's point-of-view. Topics include writing, editing, publishing, promoting and marketing books. This 328-page self-help manual is enriched with more than 100 photos, figures, illustrations, documents and publishing forms. *How To Write, Publish & Sell Your Own Aviation Books*, published by Tri-Pacer Press (1998). Contains a foreword by Walter J. Boyne, bestselling author and former director of the National Air & Space Museum (Smithsonian Institution). This 350-page self-help manual is a revised edition of *How To Get Published - Guaranteed* geared strictly towards aviation books. Numerous articles and editorials have been published in *Aviation International News*, *Centerpoint*, *The Advocate*, *The Parent Press* and *US Aviator*.
WORKS IN PROGRESS	*How To Write, Publish & Sell Your Own Nautical Books* (future publication by Tri-Pacer Press). This self-help manual will be a revised edition of *How To Get Published - Guaranteed* geared strictly towards nautical books.
FUTURE WORKS	Books for the future include *How To Promote & Market Your Own Books*, *Death Wreath*, *Avenge* and *Mountains In The Sky: True Tales Of A Sport Pilot*.
LITERARY AWARDS & HONORS	Honored Guest (along with Roxanne Pulitzer and Patrick Smith) at the 3rd Annual Palm Beach Literary Forum and Author's Luncheon in Jupiter, Florida (1989). Sponsored by the Palm Beach Community College Library. Received a handwritten letter from President George Bush stating that he had a copy of *Black Falcon* in the White House (1990). Received a Certificate of Merit for Outstanding Accomplishment from *Writer's Digest* for *How To Get Published - Guaranteed* (1994). Received a Special Book Award from the National Association of Independent Publishers for *How To Get Published - Guaranteed* (1994).
MAJOR PERSONAL APPEARANCES	New York Is Book Country, BookFest of the Palm Beaches, Miami Book Fair International, BookMania, Barnes & Noble, Books-A-Million, Bookstop, National Writers Association, AOPA Expo, Sun 'n Fun EAA Fly-In, Miami Sailboat Show.

Figure 15-8

dressed and stamped response card. Every post office in the country sells these blank postcards, which have pre-printed stamps on them. Figure 15-9 shows the front and back of the response card I sent to *Waterfront News*. As you can see, I stamped my address on the front side using an ink stamper. On the back side, I typed in several fill-in-the-blank questions, making it easy for the editor of *Waterfront News* to respond to my press kit. He didn't have to waste his own valuable time writing out a letter and stuffing it into an envelope. This method is easy and uncomplicated for everybody involved. Notice in Figure 15-9 that the editor of *Waterfront News* wrote the notation *Need Review Copy* on the bottom of his response card. This was one of the thirty publications that did not receive a review copy of *Black Falcon*, so I immediately sent one to them. The book review that was published in *Waterfront News* a couple of months later turned out to be the second best book review that was ever written about *Black Falcon*.

As a courtesy to the magazines and newspapers that send me positive response cards, I try to write each of them a thank you letter. Figure 15-10 shows the thank you letter I wrote to *Waterfront News*. Notice in this letter that I asked the editor to send me two complimentary copies of *Waterfront News*. The reason I always ask for two copies is so that I can put one copy in my vault for safe-keeping and use the other copy for display purposes when I do personal appearances. Some publications send me the complimentary copies, while others do not.

So now you see how I prepared my press kits for the first printing of *Black Falcon*. It was the first time I had ever done anything like that. If you follow my instructions, you can do the same thing with your book. It's not hard. In fact, it's a lot of fun, especially when you consider that some very famous magazine and newspaper writers might wind up reading your writing. (At least it was fun for me!) And as

America the Beautiful USA 15

TRI - PACER PRESS
P.O. BOX 840111
PEMBROKE PINES, FL 33084

© USPS 1988

We have received your Press Release about BLACK FALCON and plan to feature the book in our ____?____ issue as a book review __✓__, book excerpt_____, noteworthy news item_____. We are also interested in an interview with novelist Chevy Alden, (Yes or No).

Name John Ziegler Full job title Editor

Name of Publication Waterfront News

Address 1224 SW 1 Ave.

City/State/Zip Ft. Lauderdale, FL 33315

Need review copy

Figure 15-9

TRI-PACER PRESS, P.O. Box 840111, Pembroke Pines, Florida 33084
Publisher of Quality Action-Adventure Novels

November 3, 1988

John Ziegler, Editor
Waterfront News
1224 SW 1 Ave.
Ft. Lauderdale, FL 33315

Dear Mr. Ziegler:

Thank you for responding to our inquiry about a book review of the novel Black Falcon. A complimentary review copy will be forwarded to you immediately.

Tri-Pacer Press would appreciate two complimentary copies of Waterfront News when the book review appears in print. We would also appreciate very much if you can include the following information at the end of the review:

> Autographed copies of Black Falcon can be purchased from the publisher for $17.95 plus $1.50 for shipping. (Florida addresses add $1.08 for 6% State Tax.) Send a check or money order to: Tri-Pacer Press, P.O. Box 840111, Dept. WN, Pembroke Pines, Florida, 33084.

Once again, thank you in advance for reviewing Black Falcon.

Sincerely,

Richard Cook
Editor-in-Chief
(305)432-9683

Figure 15-10

far as the lost revenue from sending out so many complimentary review copies - well, let's just say that the free publicity I received was worth at least ten times the revenue I would have earned from selling the copies of *Black Falcon* that I gave away (probably more).

In total, nineteen publications printed some sort of write-up about my novel, which came to roughly a 25% response rate. I once heard that the average press kit brings in about a 10% response rate, so the 25% response rate that I received was excellent for a first-timer. And let me tell you, those nineteen write-ups were more than enough to fill up a large display board (which I still use as a marketing tool at air shows, boat shows and book shows). As a self-publisher, I couldn't have asked for more. But guess what? I got more! When the second printing of *Black Falcon* came off the press in 1991, I sent out a second round of press kits (this time without review copies) and ended up getting almost a dozen more write-ups in various periodicals. Figure 15-11 shows a photo that displays some of the publications that featured *Black Falcon* in either a book review or news story. The next three paragraphs are short excerpts from book reviews that were written about *Black Falcon*. These excerpts will give you an idea about why I was perfectly satisfied with the free publicity I received for my novel.

#1) *Florida Keys Keynoter*: "*Black Falcon* [is] a well-detailed journey into the world of drug smuggling and the violence that goes with it, taking readers on flights over the Keys, [with] stops at Colombian airstrips, South Florida airports and isolated Bahamian islands."[55]

#2) *Waterfront News*: "Lots of action, more than a little local color. Alden's book makes it clear that he knows his way around airplanes. He writes convincingly about flying and fliers [and] about cruising the Bahamas aboard a small sailing yacht. If you like aviation stories that remind you more than a little of Ernie Gann, you'll like *Black Falcon*."[56]

Figure 15-11

#3) *Airliners*: "The flying sequences [in] *Black Falcon* are exciting and realistic, with plenty of technical detail to show that the writer is a pilot and knows his airplanes. For aviators, the main attraction is a DC-4 and the rough-edged crews that are found around worn-out airfreighters. The author also evidently knows about sailing and the Keys from first-hand experience. The pages turn swiftly toward the inevitable showdown with the bad guys."[57]

Are you convinced yet that *Black Falcon* is a novel you should read? I sure hope so, because I'd like to sell you an autographed copy! (To order an autographed copy of *Black Falcon* either for yourself or as a gift for someone special, please refer to the *Writer's Book Mart Catalog* in the appendix section of *How To Write, Publish & Sell Your Own Aviation Books*.)

I will always be grateful to the magazines and newspapers that spotlighted *Black Falcon* during the early part of my writing career (when I was a completely unknown author). Without those write-ups, the public would never have been aware of my novel, nor would they have been aware of me. I assure you, it's a lot more fun being a known author than an unknown author. When I look back now, it seems as if one moment I was a *nobody* in the world of publishing, and the next moment I was being granted all kinds of golden opportunities, such as personal appearances, guest speaking engagements and autographing sessions (plus flights aboard many different types of aircraft).

In fact, when *Black Falcon* was originally published in 1988, I didn't know a single person who worked within the literary industry. Now, however, I personally know several hundred people who work within various literary circles, including bestselling authors, publishers, editors, bookstore owners, book show promoters, book wholesalers, book distributors and librarians. Best of all, they all know me on a first-name basis. Networking (doing business through contacts) is the lifeblood of the self-publishing industry (which happens to be a very close-knit group of people who are willing to share both information and contacts).

Networking, however, works far beyond mere literary circles. For example, networking has worked fantastically well for me within aviation and sailing circles. What I've found out is that if the networking game is played correctly, the possibilities for expansion are limitless. So far, I've been playing the game correctly. Let me now illustrate three examples of how networking has helped advance my writing career since *Black Falcon* originally came out in print. The examples I've chosen come from the three areas in which I have successfully played the networking game - the world of aviation, the world of sailing and the world of literature.

The first example of networking that I want to discuss involves the aviation industry, which happens to be my most successful avenue for promoting and marketing *Black Falcon*. This particular example of networking began in 1990 at a local airline memorabilia trade show run by the Southern

Florida Airline Historical Association - a non-profit organization that rents a meeting room semi-annually to give airline enthusiasts in South Florida the opportunity to buy, sell and trade airline timetables, airline postcards, airline playing cards, airline photographs, airline swizzle sticks, and just about any other collectible item that can be associated with airlines and airliners (including barf bags, believe it or not).

The hobby of collecting airline memorabilia is growing rapidly all over the world, especially in locations that are major airline cities, such as Los Angeles, New York, Chicago, Miami, London and Frankfurt. Many of these cities have their own local associations that sponsor small airline memorabilia shows a few times each year. These local associations are usually affiliated with the World Airline Historical Society, which sponsors its own airline memorabilia trade show every summer in a major airline city somewhere around the globe. This annual show is called Airliners International and it is held in grand fashion.

After *Black Falcon* came out in print, I attended several of the local airline memorabilia shows in the Miami area. However, because of the high cost associated with traveling to one of the international shows, I had purposely stayed away from them. It was at one of the local shows in 1990 where I received a flyer announcing the following year's Airliners International '91 memorabilia trade show, which was scheduled for July 18-21, 1991 in Orlando, Florida. Figure 15-12 shows a copy of the flyer. Orlando had only recently become a major airline city, thanks to its proximity to Busch Gardens, Disney World, Kennedy Space Center, Universal Studios and dozens of other vacation attractions located within a short distance of Orlando International Airport. I immediately sent a check to secure a booth at the Orlando event, made a hotel reservation and waited patiently for July 1991 to come around. I had no idea what was in store for me at Airliners International '91, so I hoped

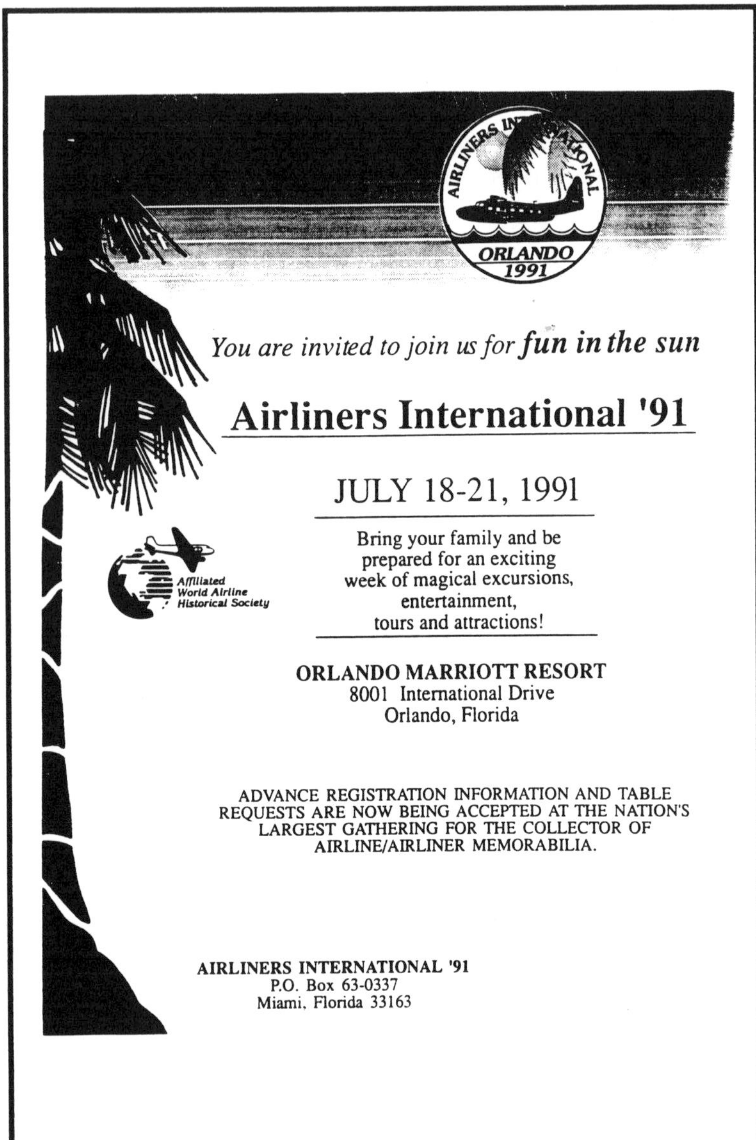
AIRLINERS INTERNATIONAL
ORLANDO
1991
You are invited to join us for fun in the sun
Airliners International '91
JULY 18-21, 1991
Bring your family and be prepared for an exciting week of magical excursions, entertainment, tours and attractions!
Affiliated World Airline Historical Society
ORLANDO MARRIOTT RESORT
8001 International Drive
Orlando, Florida
ADVANCE REGISTRATION INFORMATION AND TABLE REQUESTS ARE NOW BEING ACCEPTED AT THE NATION'S LARGEST GATHERING FOR THE COLLECTOR OF AIRLINE/AIRLINER MEMORABILIA.
AIRLINERS INTERNATIONAL '91
P.O. Box 63-0337
Miami, Florida 33163

Figure 15-12

in my heart for a successful trip.

Well, when July finally arrived I headed up towards the Orlando area, only to discover that international collectors (airline enthusiasts from every country you could possibly imagine) were more interested in purchasing barf bags and swizzle sticks than they were in buying *Black Falcon*. (Keep in mind that not every marketing idea works, but you have to at least try.) I managed to sell enough books to cover my expenses for the trip, but what really paid off was the networking I did at the show. You see, Martin Caidin was there autographing copies of his latest book *Ghosts Of The Air: True Stories Of Aerial Hauntings*. During the course of the event, I had met Martin Caidin's publicity agent, who introduced me to Caidin and his wife. I gave Caidin an autographed copy of *Black Falcon* and he gave me an autographed copy of *Ghosts Of The Air: True Stories Of Aerial Hauntings*. Figure 15-13 shows the title page from Caidin's book, on which he wrote out, "*For Chevy - Stay With It - Martin Caidin...7-20-91*". I got to talk awhile with Caidin and his wife. In fact, when I told them how I self-published *Black Falcon*, Caidin's wife seemed interested in possibly self-publishing some of her husband's out-of-print books.

In case you never heard of Martin Caidin, he was one of the most prolific aviation authors during the first century of manned-flight, having penned more than two hundred books. (Sadly, Caidin passed away on March 24, 1997.) Some of his most famous books included *Marooned* (which was turned into the classic sci-fi movie of the same name) and *Cyborg* (which was turned into the television series *The Six-Million Dollar Man*).[58] Like Ernest K. Gann (one of my mentors), Caidin had spent much of his life writing, flying and adventuring around the world - so meeting him in person and talking to him about writing aviation books was a special opportunity for me back in 1991. I had previously seen him flying his Junkers Ju-52/3m (an ex-World War II

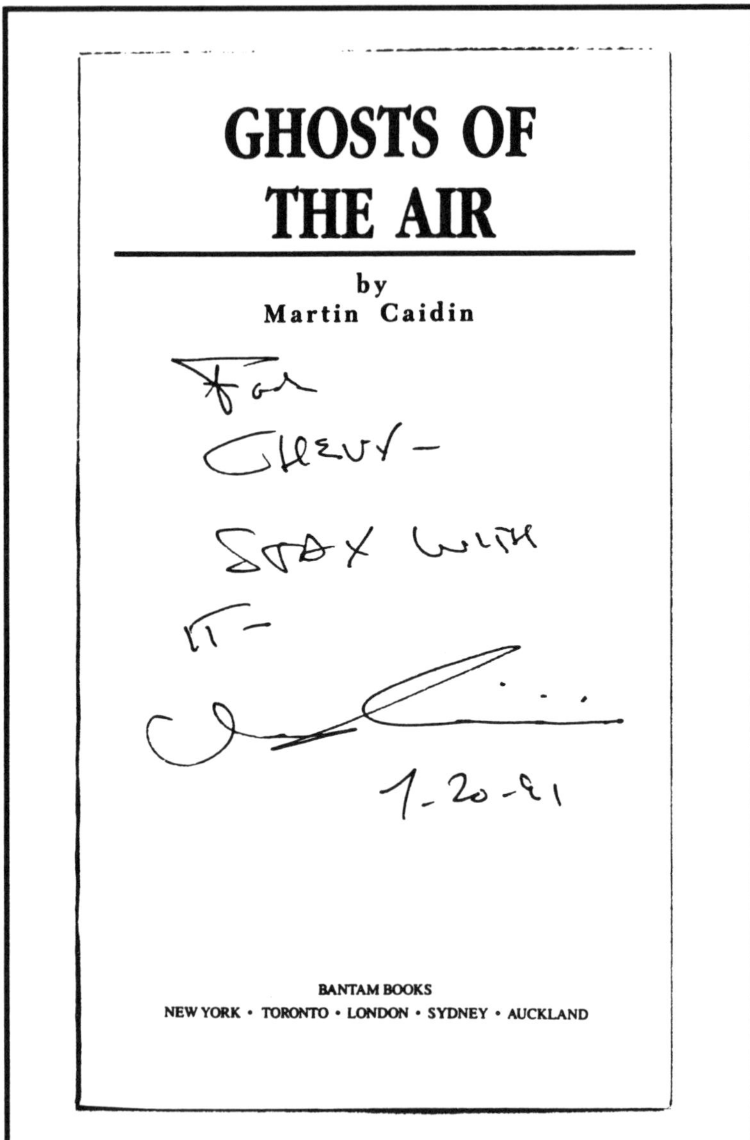

GHOSTS OF THE AIR

by
Martin Caidin

For
Chevy –
Stay with
it –

1-20-91

BANTAM BOOKS
NEW YORK • TORONTO • LONDON • SYDNEY • AUCKLAND

Figure 15-13

German transport plane affectionately known as *Iron Annie*) at various air shows (and had even refueled the plane on one occasion), but I had never actually meet Caidin in person (let alone exchange autographed books with him) until the Airliners International '91 trade show. However, meeting Martin Caidin was only the first layer of a two-tiered networking cake. The second layer came when Caidin's publicity agent told me that she was the coordinator for the After Dark Evening Program at the annual Sun 'n Fun EAA Fly-In in Lakeland, Florida - a week-long air show sponsored by the Experimental Aircraft Association as a total celebration of grassroots aviation (*a true sport pilot's utopia*). When she asked me if I would consider being a guest speaker at the 1992 Sun 'n Fun EAA Fly-In, I knew I had just been handed the entire networking layer cake on a silver platter. I accepted her invitation to speak and we set the date for the evening of April 10, 1992 - the last night of the air show. When that night came around, I made my presentation to a small crowd of approximately fifty people, telling them about my life as an aviation novelist. The reason the crowd was so small was two-fold. First, since it was the last night of the week-long event, most people had already flown home. Secondly, many of those who were still camped out at the airport (camping in a tent or trailer is one of the best ways to enjoy a grassroots air show such as Sun 'n Fun) did not want to brave the torrential downpour and high winds that had scoured the open-air building just prior to my presentation. Still, despite those two problems, my talk went extremely well.

In fact, my talk went so well that I was invited back to speak at the following year's Sun 'n Fun EAA Fly-In on the night of April 20, 1993, which was the third night of the week-long event. I had a much larger crowd the second time around - roughly 200 people. Again, my presentation went well. After my 1993 presentation, I was invited back to speak in 1994, but not during the After Dark Evening Program. Instead, I was offered the opportunity to present two daytime forums titled *How To Write And Publish Your Own Aviation Books*. You see, each day during the week-long air show, dozens of forums are given by aviation experts on all sorts of aviation-related topics. My two forums were scheduled for April 12, 1994 (the third day of the air show) and April 14, 1994 (the fifth day of the air show). Well, my two forums were so popular that the person in charge of the forums invited me back to make three presentations the following year.

I have since been back to the Sun 'n Fun EAA Fly-In every year as a guest speaker, always giving my forums to enthusiastic crowds. In 1997 (the year I was busily writing *How To Write, Publish & Sell Your Own Aviation Books*) I had a full schedule that included three forums during daylight hours, a presentation during the After Dark Evening Program, and several autographing sessions at the Sun 'n Fun Gifts of Flight store (where I was selling my books). To save on expenses during the week, I even camped out at the airport in a tent next to an EAA chapter from Georgia. How did all this happen? It happened through networking! Absolutely! Positively! Figure 15-14 shows a photo of one of the forum tents where I gave my three daytime presentations. Figure 15-15 shows a photo of the sign announcing my evening talk. Figure 15-16 shows a photo of my autographing table in the Sun 'n Fun Gifts of Flight store. My set-up in the store included a display board containing various book reviews, an airbrushed sign highlighting *Black Falcon* and, of course, lots of copies of *Black Falcon* and *How To Get Published - Guaranteed*. Figure 15-17 shows a photo of my tent and my truck at the airport campground.

Networking! That has been the key to selling my books within aviation circles. For example, at the Sun 'n Fun EAA Fly-In in 1996, I met the president of the Aircraft Owners

Figure 15-14

Figure 15-15

Figure 15-16

Figure 15-17

and Pilots Association. I told him about my aviation publishing forums and gave him one of my press kits (which I always have on hand when I do events). A couple of months later I received a telephone call from AOPA's convention and meetings director, telling me that the president of AOPA had forwarded my press kit to her and that I was scheduled to speak at AOPA Expo '96 in San Jose, California. Up until that time, all of the events I had done had taken place within the state of Florida. I knew that traveling to California would be expensive, but I didn't want to turn down such a great opportunity, so I kicked my mind into high gear and came up with a way to get to California for free - *corporate sponsorship*! And it worked! All of my expenses (including airfare, rental car, hotel and food) were paid for by a corporate sponsor (a general aviation company that ended up getting excellent exposure at my presentation). I spoke for a little less than two hours at the convention on October 18, 1996, followed by an autographing session, where I sold $900 worth of books in less than fifteen minutes. The response that AOPA received from those who attended my presentation was so overwhelming that I was invited to speak again the following year at AOPA Expo '97 in Orlando, Florida. Once again, corporate sponsorship paid my way.

And the networking goes on and on within the aviation industry! Through my various networking connections, I've gotten *Black Falcon* stocked in all types of aviation-related businesses - from aviation bookstores to fixed base operations, from aviation supply houses to air show gift shops. Figure 15-18 shows a photo of *Black Falcon* on display inside a merchandise showcase at a fixed base operation, alongside logbooks, training manuals and other pilot-related items. Figure 15-19 shows a photo of *Black Falcon* on display at the Grand Opening of a South Florida pilot shop, where I was autographing copies of my aviation novel.

Figure 15-18

Figure 15-19

Figure 15-20 shows two photographs (one over-laid upon the other) of me autographing books at another aviation event. Notice that I use this airbrushed sign at many of my aviation-related events. Figure 15-21 shows a close-up shot of this sign. Painted by a friend of a former co-worker of mine, this airbrushed sign is colorful and attracts a lot of attention, don't you think?

The next example of networking that I want to discuss with you deals with my involvement within the world of sailing (after all, many pilots are also sailors). When *Black Falcon* was first published, a press kit and review copy were sent to *Cruising World*, a popular sailing magazine geared towards cruising sailors. The cover letter was addressed to the magazine's editor. Because I never received a response card from him, I wrote off the review copy as a loss.

A few months later I read an advertisement about the Miami International Sailboat Show (which is a branch of the Miami International Boat Show), scheduled to take place in mid-February 1989 at the Miami Beach Marina. I couldn't afford to rent a booth, so I did the next best thing - I made a deal with someone who *could* afford a booth. You see, back when I was restoring my twenty-two foot sailboat, I had purchased a used bow pulpit and some marine hardware from a local boat dealer. Well, just prior to the Miami International Sailboat Show I learned that this dealer was planning to display a new trailerable, shoal-draft sailboat, which was built by the same company that built the trailerable sailboat I owned. Using this as an opportunity to network, I asked the dealer if he wanted me to talk to potential customers at his booth about the great adventures I've had aboard my own trailerable, shoal-draft sailboat in exchange for a small display table at his booth where I could sell autographed copies of *Black Falcon*. He wholeheartedly agreed to this arrangement.

Well, when the show finally came around, I got to

Figure 15-20

Figure 15-21

spend all day Friday, Saturday and Sunday at the dealer's booth talking about sailboats and selling copies of *Black Falcon*. I met a lot of interesting people, but one woman in particular caught my attention more than the rest. She kept asking me questions about *Black Falcon* and about how I published the book myself. She seemed very interested in nautical-oriented fiction. When she handed me her business card, I noticed that she was the deputy editor of *Cruising World* (and soon to become the magazine's editor). She asked me why I had never sent a review copy to *Cruising World*, so I explained that I had sent one, but that I never received a response from the editor of the magazine (her boss). She then told me that he had an extremely busy schedule and that the review copy was probably misplaced somewhere in his office. She said that she would look for it as soon as she got back to the office in Newport, Rhode Island. She also stated that she definitely wanted *Cruising World* to do a book review about *Black Falcon*. We talked a bit more, then parted company - she touring the rest of the show, while I remained at the booth.

Several months later I got a telephone call from one of the staff writers at *Cruising World*. She told me that the deputy editor of the magazine had given her a copy of *Black Falcon* and that she was writing a brief book review about my novel for inclusion in an article about nautical books. She said that everyone on the *Cruising World* staff who had read my book really enjoyed it and that the reason she was calling was to ask me some personal questions. I answered the questions the best I could, then waited around for the article to appear in the magazine.

Low and behold, when the June 1989 issue of *Cruising World* came out, there was my book proudly displayed in a photo featuring a dozen or so nautical books and videos. The title of the article was *Summer Editions: Books And Videos To Tide You Over When Storm Warnings Fly And*

Your Boat Can't. Within the article itself was a very good review about *Black Falcon*, along with the address for Tri-Pacer Press and the retail price of the book. Mail-order sales from sailors soon started coming in, plus a couple of commercial orders from nautical bookstores, plus one personal appearance at a Fort Lauderdale marine supply store. Briefly, here is how this true example of networking went. A personal contact at a boat dealership led to a personal appearance at a sailboat show, which led to a personal contact at a sailing magazine, which led to other personal contacts at that sailing magazine, which led to a book review in that magazine, which led to retail sales, commercial sales and a personal appearance at a marine store. True networking in action!

The last example of networking that I want to discuss with you revolves around the world of literature and books. This example started out as a simple press release, which was mailed to approximately 200 Florida-based libraries. I wanted to make them aware of *Black Falcon* - a novel that featured Florida locales throughout the story line. You see, I learned that most libraries had special sections for books that featured local interests, as well as books written by local authors. I figured I'd try to sell *Black Falcon* to at least a few Florida-based libraries since my novel fit both categories (local interest and local author).

Well, quite a few orders came in as a result of that press release. More importantly, though, was the letter I received from the director of the Palm Beach Community College North Campus Library. She stated in her letter that after reading my press release, she felt I was qualified to be a guest author at the 3rd Annual Literary Forum & Author's Luncheon in Palm Beach County on May 25, 1989. Other local authors who were scheduled to appear included F. Lee Bailey, James Hall, Roxanne Pulitzer and Patrick Smith. Imagine the elation I felt over the chance to mingle with those famous authors. Naturally, I accepted the invitation.

On the 24th of May, I was a blue-collar worker at Fort Lauderdale-Hollywood International Airport, spending the day refueling aircraft. On the 25th of May, I was suddenly thrust into the world of high-society at the Admiral's Cove Clubhouse in Jupiter, Florida, where I was one of the authors being honored by people who came to eat (or should I say *dine*) with us. It was there that I realized how it felt to be treated like a real celebrity. In fact, I was treated as if I was made out of gold and diamonds! It was the most incredible feeling - almost as if I couldn't do or say anything wrong. For several hours I lived the life of the rich and famous and, although it was wonderful and exciting, it was also kind of eerie. I mean, what if every day was like that? It would be impossible to live a normal lifestyle. I guess that's why so many celebrities live the way they do. It's sort of forced upon them by their admiring public.

Figure 15-22 shows the program from that luncheon, which included a list of the guest authors. Notice how my name appeared at the top of the list. Was I smart when I chose my pseudonym or what? You bet I was! The name Chevy Alden was chosen for a specific reason and it has paid off time after time.

Overall, the event was a great success and I will always be grateful to the director of the Palm Beach Community College North Campus Library for taking a chance on me when I was just an unknown novelist from South Florida. At one point during the luncheon, she stood up to individually introduce each author to the people at the gathering, telling a little bit about each author's latest book. When she got to me, she told everybody about *Black Falcon*, then said that one of my future goals was to go long-distance cruising aboard my own sailboat. A woman in attendance asked out loud if she could go with me. To say the least, I felt like a movie star. The reception I received from that

3rd Annual Literary Forum
Author's Luncheon

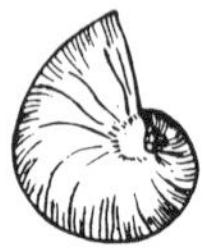

May 25, 1989

Honored Guests

Chevy Alden
Adelaide Altman
F. Lee Bailey
Joella Cain
Sydney Clary
Linda Marx Cole
Donald Curl
Sue Dauphin
Edith Delatush
Nicole Duplaix
Barbara Gibson
James Hall
C. Colburn Hardy
The Honorable James Knott
Sharon Komlos
Bruce Nash
Nick Nixon
Roxanne Pulitzer
Patrick Smith
Dr. Barry Russal
James Warnke
Bob Weller
Donald Wenberg

Admirals Cove Clubhouse
Jupiter, Florida

Authors Reception
Special Guest: F. Lee Bailey

Autographing and sale of books ~ 10:30am

Luncheon
Noon

Program

Welcome .... Dr. Ottis Smith, Provost, PBCC North Campus

Invocation .... The Reverend Frederick LeCrone, Grace Episcopal Church

Introduction of Authors .... Alice Zacherl, Director, PBCC North Campus Library

Introduction of Speaker .... Ms. Leah Schad, Chairperson, Board of Florida Audubon Society

Patrick D. Smith

Closing Remarks .... Dr. Edward M. Eissey, President, Palm Beach Community College

Figure 15-22

crowd was just unbelievable!

Then, on the following day, I was back at the airport living my blue-collar existence. It was a big emotional letdown for me, even though I enjoyed my job at the airport - a job I had been doing since I was seventeen years old. But at least I had my day in the limelight - a day that will always remain one of the highlights of my writing career.

About a year after my participation at the 3rd Annual Literary Forum & Author's Luncheon, I received a letter from somebody who had been in attendance at that event. This person belonged to a group of Palm Beach authors (some published, some unpublished) known collectively as the Backroom Writers. The purpose of that group was to meet once a month to discuss each member's writing and to give each other encouragement, support and constructive criticism. In the letter, I was asked to be a guest speaker at one of their monthly meetings. I accepted the invitation and spoke to the group on October 29, 1990. It was at that meeting that the idea for my second book was tossed out at me (which put the sequel to *Black Falcon* on hold). I had just finished discussing how I had self-published *Black Falcon* when someone shouted that I should write down all of my experiences and put them in a book. Everyone in attendance agreed that a book like that would easily be a good seller. And thus *How To Get Published - Guaranteed* was born, which went on to win an award from the National Association of Independent Publishers in 1994.

During February 1991, I received another letter - this time from somebody who had been in attendance at the Backroom Writers group meeting on the night I had spoken. In the letter, I was asked to participate in a panel discussion titled *How To Write A Book And Get It Published*, which was going to take place at the very first BookFest of the Palm Beaches, scheduled for April 19-21, 1991. Well, not only did I accept the invitation to be a guest panelist, but I also became a book vendor. Tri-Pacer Press had its own booth during the entire three-day literary event - an event that has since become an annual affair for me and which opened up a whole new marketing avenue for my books. Up until that time, I had never been to a book show, let alone be a vendor at one.

The networking contacts made possible by my participation at the BookFest of the Palm Beaches were beyond my wildest dreams and too numerous to mention here. One of those contacts, however, hooked me up with the right connection so that I was able to become a book vendor at that year's Miami Book Fair International (the largest street-type book festival in the world, which is held each November on the weekend before Thanksgiving).

The BookFest of the Palm Beaches and the Miami Book Fair International have become annual events for me. Both are real money-makers for Tri-Pacer Press and both lead to new networking possibilities every year. I have also done smaller book shows throughout Florida, but I don't necessarily attend these other shows annually (depending on what else is going on at the time). Figure 15-23 shows a photo of my booth at the Miami Book Fair International in 1991, when *Black Falcon* was the only book I was selling. Notice the large display board that I was using at the time with all of the book reviews and articles mounted on it. I have since changed this board to the smaller one shown in Figure 15-16 a few pages back. Figure 15-24 shows a photo of my booth at the BookFest of the Palm Beaches in 1993, when *How To Get Published - Guaranteed* was still being written. Notice that I was displaying an enlarged pre-publication mock-up of the front cover of *How To Get Published - Guaranteed* (which was eventually changed before going to press). I was using the mock-up to bring in advanced orders for the book. Notice, too, that another self-publisher was sharing my booth. The reason I started sharing my ven-

Figure 15-23

Figure 15-24

dor booth was to cut my expenses in half. That year there were half-a-dozen self-publishers at the BookFest of the Palm Beaches - a sure sign that more and more people were choosing to guarantee the publication of their books.

Book shows are tiring, but they are also a lot of fun. The networking possibilities are limitless, providing self-publishers with the opportunity to meet authors, publishers, agents, bookstore owners and other literary-minded people. As I progress forward with Tri-Pacer Press, I plan on participating in more and more book shows all across the nation. In fact, I finally made it to the *Big Apple* in 1997. That's right - New York City! On September 21, 1997, I had a vendor booth at the New York Is Book Country book show on Fifth Avenue in New York City. (Coincidentally, this was also the 10th Anniversary of my decision to self-publish *Black Falcon* after receiving eighty-four rejections from New York publishers.) So many events, so little time - but I will find a way. And networking will play a big role in it all.

Everything I have discussed within this chapter boils down to one goal - selling copies of your book. Without this as your ultimate goal, there's no point spending time promoting and marketing your book. So, naturally, I'm assuming that you *do* want to sell copies of your book. I'd like to give you one more example of how to promote and market your book before I end this very important chapter. Please pay attention because this example can be one of your most useful sales tools if you do it correctly. I'm referring to business cards. Simple, old-fashioned business cards. Don't cut costs when designing these little pieces of paper. Spend as much time and money as you have to so that your business cards look professional. Also, have them printed by a print shop - not on a home computer (like I've seen a lot of people do). I have two-sided business cards that use one side to advertise my speaking engagements and the other side to advertise my books. I carry these cards everywhere I go and

I pass them out liberally. Rather than explaining all of the information that is contained on my business cards, take a look at both sides for yourself in Figures 15-25 and 15-26. Notice how the drawings of the sailboat and the airplane attract your attention. These business cards really work!

You have now seen that the possibilities for selling self-published books are almost endless - limited only by an author's time, effort and ingenuity. In order for an author to assure the success of his or her book, he or she must do two things. First of all, he or she *must write a good book*. An excellent manual that can help an author get started with this task is titled *Is There A Book Inside You?*. Secondly, once an author self-publishes his or her book, he or she *must promote and market that book appropriately, according to the type of book that it is*. An excellent manual that can help a self-publisher with this task is titled *1001 Ways To Market Your Books*. (To order a copy of either of these manuals, please refer to the *Writer's Book Mart Catalog* in the appendix section of *How To Write, Publish & Sell Your Own Aviation Books*.)

My fourth book, tentatively titled *How To Promote & Market Your Own Books*, should be available by the year 2000, so contact Tri-Pacer Press at the turn-of-the-century for ordering information. (Is that a cool sentence or what? The turn-of-the-century!) This self-help manual will delve deeply into every aspect of my unique way of promoting and marketing books. In other words, it will be a greatly expanded version of this chapter. If you've enjoyed reading Chapter Fifteen and you want to learn more, then *How To Promote & Market Your Own Books* will definitely be a book you'll want to read. This self-help manual will contain priceless information, such as how to arrange personal appearances and speaking engagements, how to arrange corporate sponsorship deals, how to find someone to write the foreword to your book, how to get libraries, bookstores,

CHEVY ALDEN
Guest Speaker

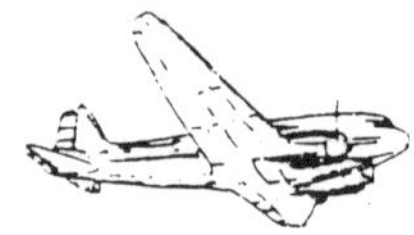

Author, Pilot, Sailor, Adventurer, Thrill-Seeker

Alden Speaking Engagements, c/o Tri-Pacer Press
P.O. Box 840111, Pembroke Pines, FL 33084-2111
Phone: (954) 432-9683 Fax: (305) 652-6296

Figure 15-25

Books by Chevy Alden:

How To Get Published-Guaranteed (328 Pages) - This self-help manual is a "must-read" for every author who wants to learn about publishing from an insider's point-of-view. Topics include writing, editing, publishing, promotion and marketing. ($18.95)

Black Falcon (257 Pages) - A thrilling action-adventure novel about drug smuggling that takes readers from a sleazy Miami bar to a salty Key West tavern, from the cockpit of a drug plane to a clandestine airfield in the mountains of Colombia, from a tropical Bahamian cruise aboard a private sailboat to a climactic action-packed ending on Fort Lauderdale's Venice-like canal system. ($18.95)

Shipping - Add $2.00 for first book, $0.75 for each add'l book.
Sales Tax - Florida addresses add 6% State Sales Tax.

Send check or money order to: Tri-Pacer Press, P.O. Box 840111
Pembroke Pines, FL 33084-2111

Figure 15-26

book wholesalers and book distributors to buy your book, plus a whole lot more.

At what point can a self-publisher be considered a success? Well, in my opinion, when a self-publisher sells 1,000 copies of his or her book, the venture should be considered a success. At that point, all future sales can be considered gravy. Why do I feel this way? Because a self-publisher who thinks that he or she will sell 5,000 copies, 10,000 copies, 50,000 copies, or even 100,000 copies within a very short time span is bound to be disappointed when he or she sells anything less than that amount. Sure, it's possible to sell that many copies of a self-published book, but it's more common to sell about 1,000 to 5,000 copies (which is still excellent). Heck, even major publishers have a difficult time selling more than 5,000 copies of a book written by an unknown author, so be happy with 1,000 copies as an initial goal for success.

But what if a self-publisher can't sell even 1,000 copies of his or her book? What does that mean? Well, it either means that the author did not write a good book or the book was not manufactured professionally or the promotion and marketing efforts were not carried out correctly. The important thing is to at least give it a try, because without trying an author will never know whether or not his or her book could have been a success.

The fantastic thing about promoting and marketing a self-published book is that everything is in the author's control. Nobody else can take the blame for failure, nor can anyone else claim the credit for success.

CHAPTER SIXTEEN

A PERSONAL MESSAGE FROM CHEVY ALDEN

Now that you know the true story behind the making of *Black Falcon* (as well as the making of my writing career), you should be in agreement that it's definitely possible for an unknown and unpublished author to *guarantee* the publication of his or her own book. It doesn't matter who the author is, as long as he or she possesses the drive and determination to start their self-publishing venture as a long-term endeavor.

In *How To Write, Publish & Sell Your Own Aviation Books*, I told you about the many literary adventures I experienced during my first five years as an author and publisher. I wouldn't trade those years for anything because they allowed me to learn the intricate workings of the publishing industry from an insider's point-of-view. I started out as a one-book publisher and slowly progressed forward to the point where I was ready for expansion. That expansion took place during the beginning of my sixth year as a self-publisher in the form of a new division of Tri-Pacer Press called *Writer's Book Mart Catalog*. This division acts as a clearinghouse for authors who are truly interested in self-publishing their own books. As such, I supply these authors with the same reference manuals I have been using in my daily dealings as a writer and self-publisher - thereby saving them the time of tracking down these valuable reference books. (For further information about *Writer's Book Mart Catalog* and the publications it supplies, please refer to the appendix section of *How To Write, Publish & Sell Your Own Aviation Books*.)

Ironically, *Writer's Book Mart Catalog* was created in conjunction with the publication of my second book, *How To Get Published - Guaranteed*, in 1994. Together, the success of these two ventures (along with continued sales of *Black Falcon*) allowed me to quit my part-time job as a high school substitute teacher - 1996 having been my last year in a classroom. Hopefully, the future will bring financial independence through my various publishing efforts, allowing me to quit my full-time job as an aircraft refueler. The publication of my third book (the one you are now reading) in 1998 definitely brought me a little closer to this goal - and I have more than half-a-dozen books outlined, so the future looks very bright for me if I continue to do as well as I have in the past.

Now that I've reached my original goal of emulating the lifestyle of Ernest K. Gann as a pilot, sailor and author, I've had to come up with new goals. Let me tell you a little bit about them. First of all, I'd like to buy my own airplane someday (possibly a classic Piper Tri-Pacer, possibly something else) and fly it around the continental United States. Next, I'd like to sell my twenty-two foot sailboat (the one I restored) and buy something in the thirty foot range, then spend a couple of years cruising down to the Caribbean area (flying home occasionally to take care of publishing-related business). I would also like to relocate my home and business to waterfront property so that I can keep my sailboat in the backyard. The only stipulation is that an airport must be no further than a fifteen minute drive from my property. That way, I can fulfill my ultimate goal, which is to wake up each morning and ask myself, "What should I do today - write, fly or sail?" Is there any question as to why my current sailboat is named *Dreamer*? I have dreams. *Real big dreams!* That's why when I autograph books, I usually sign the inscription *Live Your Dreams*.

I have now told you how I became a successfully published author. Now you must make the same decision I was faced with back in 1987 - to self-publish or not to self-publish. *Remember, only you can guarantee the publication of your book!*

To help you make the right choice, let me offer a few suggestions. First of all, read *How To Write, Publish & Sell Your Own Aviation Books* again from beginning to end. Next, think about your long-term goals, because if you want self-publishing to pay off, it has to be a long-term venture. And finally, ask yourself if you want to be in control of your own writing career *right now* or if you want to wait around for somebody else to possibly control it for you (which might never happen). If your answer is to sit around and wait for somebody else, then self-publishing is definitely not for you. On the other hand, if you want to be in control of your own destiny (or even if you are still undecided), then self-publishing just might be the right avenue for you to take.

Does self-publishing work? The answer is yes! I get first edition copies of books all the time from authors who have successfully self-published their own books after following my advice. Figure 16-1 shows a letter I received from Leslie Robinet (a total stranger up until the time she sent me the letter) telling me that she and her husband, Bill, had started Billville Press in Veneta, Oregon, after reading *How To Get Published - Guaranteed*. They were self-publishing Bill's autobiography about his years as a cropduster during the 1950s and wanted to know if I would write a cover blurb for the back of their dustjacket. They sent me a copy of Bill's manuscript (along with some photos that would appear in the book) and I loved it. Figure 16-2 shows the letter I sent back to them containing the cover blurb. I wrote, "A fascinating, true-grit account of ag flying during

BILLVILLE PRESS

Veneta, Or. 97487

March 16, 1997

Tri-Pacer Press
P.O. Box 840111
Pembroke Pines, Fl. 33084
Fax (954) 652-6296

ATTN: Chevy Alden

Dear Chevy (Richard),

I have your self-publishing book and it has been very valuable with my on-going project. My husband has written a book about his experiences as a cropduster in the 1950's flying Stearmans, N3N's and Travel Airs.

We are preparing to publish this book and a friend of mine met you last year at the AOPA convention and mentioned it to you. You gave her your card and she indicated that you may be available to give us a blurb for the back of the dust jacket. I think that the author of an aviation novel would have good credibility on our dust jacket.

I'm faxing this note because of the time difference and I'm away from the phone a lot. If you would consider reviewing some or all of the book, please fax me a note and I'll get the manuscript to you. We are planning to go the printer in about two weeks. The cover art is the hold-up right now.

Thanks to your book I was able to do the typesetting myself with my computer. Hope to hear from you soon.

Sincerely,

Leslie Robinet

Phone:
Fax:

Figure 16-1

TRI-PACER PRESS, P.O. Box 840111, Pembroke Pines, Florida 33084-2111 • 954-432-9683

Publisher of Quality Titles Since 1988

March 24, 1997

Bill and Leslie Robinet
Billville Press
Veneta, OR 97487

Dear Bill and Leslie:

By The Skin Of My Teeth: A Cropduster's Story is fabulous. I started reading it over the weekend while camping with my daughter in the Everglades. What a fun read it is! Here is the blurb you requested for the back of the dustjacket:

> "A fascinating, true-grit account of ag flying during the '50s. If you like biplanes and low-level flying, you'll love Robinet's memoirs."
>
> Chevy Alden - Award-Winning Author of *How To Get Published - Guaranteed*

My favorite line from the book is, "It isn't enough to kill the damned bugs, but more important to scare and torture the hell out of them first, while holding a private airshow for the customer." I've read *Biplane* and *Nothing By Chance* by Richard Bach (both of which are books about flying biplanes), so I can't wait to add a first edition copy of *By The Skin Of My Teeth: A Cropduster's Story* to my aviation book collection.

As a publisher, I must point out one mistake that you must fix, otherwise every book reviewer will take notice immediately and it will be a big strike against your publishing company. Many self-publishers make this mistake. Foreword is not spelled Forward. Please change this immediately before your book goes to press. Other than that, you have done an excellent job with the book so far and I am proud that *How To Get Published - Guaranteed* had something to do with it.

If you can make it down to the Sun 'n Fun EAA Fly-In (2nd only to Oshkosh) this year, I will be giving three forums (Sunday, Wednesday and Saturday) during the week of April 6 - 12, plus I will be the featured speaker during one of the Evening Programs (Tuesday). I will also be selling my books in the Gifts Of Flight store throughout the entire week of the airshow.

I hope I have provided you with the kind of blurb you were hoping for. It was written with all sincerity.

Sincerely,

Chevy Alden

Figure 16-2

the '50s. If you like biplanes and low-level flying, you'll love Robinet's memoirs." A couple of months later I received a hardcover copy of *By The Skin Of My Teeth: A Cropduster's Story*. A few months after that I spotted an excellent review about Robinet's book in an issue of *General Aviation News & Flyer*.

Yes, self-publishing works! And yes, I love receiving letters from people who have successfully followed my advice. Over the years, I have received countless letters from people who have either read one of my books or attended one of my talks. These letters are always heart-warming, which let's me know that the sacrifices I've made throughout the years have all been well worth it - much more so than the money I've earned from my writing and publishing endeavors. Figures 16-3, 16-4 and 16-5 show a sampling of the types of letters I've received in regards to *Black Falcon*, *How To Get Published - Guaranteed* and my speaking engagements. Absolutely nothing brings me more joy than receiving letters like the ones shown here.

Before I conclude the text portion of this book, I'd like to leave you with one final example of real-life inspiration. This true story revolves around a book titled *The World Record Paper Airplane Book* by Ken Blackburn and Jeff Lammers. Jeff Lammers and I had never met prior to the 1997 Sun 'n Fun EAA Fly-In, where we had a coincidental introduction while he was setting up for a stint at the autographing table in the Sun 'n Fun Gifts of Flight store.

Lammers had a background as a mechanical engineer, designing rockets and robots for a major aeronautical firm. One of his colleagues, Ken Blackburn, had designed a paper airplane that held the Guinness World Record for flight duration of a paper aircraft. Lammers had also been a designer of paper airplanes. Together, they came up with the idea for a how-to book about making paper airplanes, using Blackburn's winning design as the title of the book. When they

AIR AMBULANCE
ASSOCIATES

WHEN LIFE ITSELF IS THE FLIGHT PLAN ... ONLY THE BEST WILL DO

FT LAUDERDALE INTNL AIRPORT
FT LAUDERDALE, FL 33315
IN FL TOLL FREE
OUTSIDE FL & USA
TELEX

November 17, 1988

Mr. Chevy Alden

Dear Chevy:

As you know, I purchased a copy of your novel BLACK FALCON yesterday afternoon. I began reading it last night about 7:30. I finished it about 01:30 this morning. Simply stated, I couldn't put it down.

As I am an avid reader, but with little spare time, a book must capture my attention within the first two or three chapters or I will discard it and begin another. BLACK FALCON hooked me somewhere around the middle of page one.

Your writing style is truly unique. I hope you never change it. I can't tell you how many books I've trashed prior to completion because the author felt obliged to add pages by inserting trivial dialogue. Your absence of this was refreshing.

I also commend you on your research. Having flown Lear Jets throughout the Caribbean, Central and South America on aeromedical missions, I can attest to the authenticity of your locales and other technical information.

Congratulations on a job well done. I believe your future as a writer is very bright. Please consider this letter an advance order for DEATH WREATH. I'm looking forward to it's publication.

Sincerely,

Walter A. Muharsky
Vice President/General Manager
Airline Transport Pilot Examiner

Figure 16-3

8-15-94

DEAR Chevy,

JUST WANTED TO LET you know that I Received "HTGP-G" ON Aug. 1 (OR somewhere close to that date). I read most of it deadheading to work and on layovers. I thoroughly enjoyed it & found it full of information in getting published. — Now all I have to do is write a book!! I have a couple of things in mind — so am ORDERING 2 books from you to — hopefully — help me along.

As you ARE fond of TRI-PACERS — my preference are Pitts. (could you tell?). My neighbor owns a TRI-PACER & I ended up flying it to OSHKOSH last month as my Pitts had a mag problem. Hope you get to fly a Pitts someday if you have any wish to — After spending 9 hrs. in the TRI-PACER — I'll take the Pitts any day!!

Thanks

Greg [illegible]

Figure 16-4

DEBBIE BRVENIK

Coconut Creek, Florida 33073

September 22, 1996

Chevy Alden
P.O. Box
Pembroke Pines, FL

Dear Chevy,

I came to hear you speak at Barnes & Noble yesterday. I wasn't really expecting much - just another bigshot writer trying to sell books. Boy was I surprised to find you there - Just an ordinary working person like me!

To write a book and get it published has been a lifelong dream of mine, but it has always seemed so out of reach that I've never been motivated enough to get through even *one full chapter*! Hearing your story has made me realize that I don't have to worry about spending possibly years to write something that might get rejected, but rather that I can take control and *make* it successful.

Last week I started the Anthony Robbins 30 day seminar on tape, to try and get myself motivated and change my life, but I think that listening to you speak yesterday has given me more drive and confidence than all $200.00 worth of Tony Robbins' tapes!! Naturally I bought your book, and have already started reading it. I also started a journal today, and have made a commitment to myself to write in it every day.

Thank you for giving me that push that I needed. You've obviously worked hard to live your dreams and I wish you all the success you deserve.

With Kind Regards,

Debbie Brvenik

Figure 16-5

failed to find a publisher, Lammers decided to become an entrepreneur by self-publishing the book himself. He called his company World Record Paper Airplanes and came out with his first printing in 1991, which included detailed instructions for making paper airplanes, plus thirty sheets of heavy stock paper for making the airplanes.

Rather than targeting pilots, Lammers marketed his book at toy and hobby shows. He certainly knew what he was doing, because he ended up selling about 20,000 copies in two-and-a-half years. He then hit the self-publishing jackpot in 1994 when his book was sold to a New York publishing company. The contract called for a 7% royalty on the first 50,000 copies and 8% on every copy thereafter. As of the time I had interviewed Lammers at the Sun 'n Fun EAA Fly-In, approximately 660,000 copies of *The World Record Paper Airplane Book* had been sold. Lammers told me that he had fully retired from his career as a mechanical engineer due to the royalties he had earned from his book and that he was spending his days working on a Mustang II kitplane. At the time, he was only thirty-five years old. Not bad for a writer who started out as a self-publisher, wouldn't you say?

Well, it is now time to bring *How To Write, Publish & Sell Your Own Aviation Books* to an end. I hope I've given you a bit of insight and inspiration as far as self-publishing is concerned. Although I now consider myself to be somewhat of a semi-expert in the field of self-publishing, my beginnings were as a novice writer who knew absolutely nothing about the publishing trade. After learning about self-publishing in 1987, I decided to give it a try. Now the choice is yours. And with that mentioned, the only thing left for me to say is, "Good-bye and good luck!"

(Please see next page for notes)

Note #1: While typesetting *How To Write, Publish & Sell Your Own Aviation Books*, I met one of the small press managers from the corporate offices of Barnes & Noble, the largest bookstore chain in the United States. Due to that encounter, Barnes & Noble decided to put *How To Get Published - Guaranteed* into their national database system and stock it as a standard item at many of their locations across the country, which required me to go back to press for a second printing of the book. Read how this all came about in *How To Promote & Market Your Own Books*.

Note #2: For further insight about writing and publishing your own books, be sure to read *Birth Of A Book* in the appendix section of *How To Write, Publish & Sell Your Own Aviation Books*.

Note #3: For a sneak preview of my action-adventure novel, *Black Falcon*, be sure to read the excerpts I have included in the appendix section of *How To Write, Publish & Sell Your Own Aviation Books*.

Note #4: If you know anyone who is looking for a highly-motivated guest speaker, tell them I'm always available for speaking engagements. They can contact me at Tri-Pacer Press, POB 840111, Pembroke Pines, FL 33084-2111. My phone number is 954-432-9683. My fax is 305-652-6296.

Note #5: If you decide to self-publish your writing, please send me a complimentary review copy of your book. I talk to a lot of aviation-oriented people and one of the greatest sales tools you can have on your side is word-of-mouth.

NOTES

(This space is provided for your own personal notes.)

APPENDIX A

BIRTH OF A BOOK

Midway through 1990 I decided to start writing a personal journal for the sole purpose of analyzing my thoughts, feelings, emotions and dreams. It didn't take long, however, before my journal became an examination of my entire life (past, present and future). From the time I started writing my journal on June 14, 1990, until the day I finished writing the first draft of *How To Write, Publish & Sell Your Own Aviation Books* on June 19, 1997, almost 1,700 handwritten pages of journal notes had been entered into half-a-dozen spiral notebooks (single-spaced, college-ruled, 81/2" x 11"). Included in those notes were comments about the creation of my award-winning book *How To Get Published - Guaranteed.*

The following excerpts were pieced together from my actual journal notes to give you a real-life look at how a book is created from start to finish, including all of the ups and downs and joys and frustrations that go along with the writing life. I call this assorted collection of excerpts *Birth Of A Book.* If you have ever dreamed about writing a book (or even if you've already written one), then you will find these passages fascinating. A few unrelated passages have also been included here to give you a brief glimpse of my personal life. Please keep in mind that while I was writing *How To Get Published - Guaranteed,* I was also working at a full-time job, part-time job, spending time promoting and marketing *Black Falcon,* spending time with my family, spending time doing home improvement chores, plus writing hundreds of pages of journal entries during the course of those several years. (Note: The following journal excerpts are unedited.)

June 14, 1990 (pg.1) - My 32nd birthday was 24 days ago, so why am I starting this journal at this stage of my life? Simply because I thought up the idea today - not yesterday or last week or last year or when I was a lot younger, although I wish I had.

October 29, 1990 (pg. 103) - I gave a speech tonight up in Palm Beach to the Backroom Writers group. The talk went really well. Everyone was very attentive and asked some good questions. They all thought that a book based on my speech would sell well - about how I beat the odds and got published. I definitely need to start working on this type of book.

November 22, 1990 (pg. 115) - I would like to write my book about self-publishing if I ever get the time. It will be titled "How To Get Published - Guaranteed".

March 21, 1991 (pg. 214) - The second printing of "Black Falcon" is now officially on the market. The books arrived a day early.

April 22, 1991 (pg. 232) - It was an eventful three days at BookFest. On Sunday from 10:30am to 12:30pm I participated in the "How To Write And Publish A Book" panel discussion. I found out from this panel discussion that my book "How To Get Published - Guaranteed" is going to be a big seller. About 200 people attended and most of them were frustrated, unpublished authors. I definitely have to find the time to get seriously working on my book.

May 1, 1991 (pg. 233) - I haven't had a chance to write in this journal in over a week-and-a-half. I started writing "How To Get Published - Guaranteed" on April 28th. I was at work and the ideas just started flowing in my head for various chapters, sections, paragraphs - and the ideas haven't stopped flowing yet, so I'm constantly writing notes on scraps of paper while driving, lying in bed, everywhere. It's just like when I first started to write "Black Falcon". I haven't had time yet at night to sit down and actually start writing page by page. I wonder how long it will take me to finish this book? Even though this book is barely started, I've been receiving mail in reference to it. I got a letter yesterday from a pilot who flew bush planes up in Alaska. He wants to write and publish a book about it. He heard my comments at the BookFest panel discussion and found my talk interesting.

May 21, 1991 (pg. 245) - Two envelopes arrived in my post office box yesterday. One was from the BookFest Author's Committee. Enclosed was a $100 check for my participation in the panel discussion on "How To Write And Publish A Book". This check was totally unexpected and it is the first time I've ever been paid an author's fee for talking to people about my writing. The other envelope was from a local marine supply store called Sailorman. This store puts out a bimonthly magazine for sailors called "Sailorman Star". They are going to do a book review about "Black Falcon" in the June/July 1991 issue. They want to stock the book in case they have people asking for it.

May 22, 1991 (pg. 246) - I turned 33 years old today. It was a damp, rainy, soggy kind of a birthday. I was suppose to have gone flying, but a tropical depression moved in last night and it's been solid IFR weather all day. Low ceilings, high winds with higher gusts, not much visibility, lots of rainshowers and thunderstorms. I had to cancel the flight.

June 4, 1991 (pg. 256) - The book review on "Black Falcon" just came out in the June/July 1991 issue of "Sailorman Star". It's a good one - very positive.

June 6, 1991 (pg. 259) - Today was my last day on the day shift at the airport. I'm glad. I've always thought airports were more mysterious at night. More romantic. More adventurous. Most of the flying scenes in "Black Falcon" take place at night for just this reason. The night shift is now mine from 4:00pm until 12:00am, Sunday through Thursday!

June 14, 1991 (pg. 263) - This is a very special day! It was exactly one year ago this morning when I came up with the idea for this journal. Never did I think it would have come this far - 263 pages of thoughts, dreams, experiences, hopes, fears, revelations. This journal is truly turning out to be my life on paper. I'll have to see what path this journal takes during the next 366 days (leap year 1992).

June 19, 1991 (pg. 264) - Just when everything seems to be going smoothly, something always has to @$#& [*4-letter word deleted*] up. Yesterday our roof sprung a leak during a heavy thunderstorm.

July 5, 1991 (pg. 276) - I got a card today from "South Florida" magazine. They are doing an article about "Black Falcon" in their August 1991 issue.

July 15, 1991 (pg. 276) - The August issue of "South Florida" magazine came out already. They did a quarter of a page devoted to four books that have local interest - "Black Falcon" being one of them. There is a brief review of each book and pictures of two of the books - "Black Falcon" being one of them. And the picture is in color. This is a first. The August issue is a fitting one to have "Black Falcon" in because there are feature articles about the Everglades and Key West - both of which are in my book.

July 29, 1991 (pg. 312) - Yes! Yes! Yes! I finally did it! I got into the Miami Book Fair International for this coming fall on November 15th, 16th and 17th. It's one of the largest street book fairs geared towards readers as opposed to librarians or booksellers.

October 1, 1991 (pg. 343) - It was in October 1981 when the idea for the novel "Falcon" (now "Black Falcon") popped into my head. I don't remember the exact date, so I always consider October 1st as the Anniversary date. So it's now 10 years since I made the decision -

against all odds - to write a novel. And look how far I've come! It's a REAL accomplishment. Where will I be in another 10 years? Alive? Dead? Flying cross-country? Sailing the Caribbean? 3 books under my belt? Who knows? I was cleaning out my sailboat today to get it ready for next week's sailing regatta. I discovered a lot of water in the bilge. I traced the leaks to the 3 portholes. So I took all the gear out of the boat and I now have to remove the frames and reseal them. The boat is clean, though. It was an all-day job. I also flushed out a wasp's nest that was under the bowsprit. 5 large wasps flew out. I hosed them away.

October 15, 1991 (pg. 360) - What a week! Writing, sailing, flying, sex! Life doesn't get much better than this. If I didn't have to work, it would be perfect.

November 18, 1991 (pg. 384) - The Miami Book Fair International is over. It was not the success I had hoped it would be. In fact, it was downright disappointing.

November 20, 1991 (pg. 389) - I've scheduled a plane for next Tuesday. I can't wait for that day to come around. I need some solitude up in the air. I think I'll take a flight up to the farms south of Lake Okeechobee.

December 7, 1991 (pg. 398) - I've put "How To Get Published - Guaranteed" on hold. Is there a pattern emerging?

December 9, 1991 (pg. 399) - Maybe I should be satisfied just being the blue-collar worker that I am, and with writing this journal, and flying, and sailing. Maybe I don't even have other books inside of me the way I had "Black Falcon" inside of me. Maybe all I have is other book ideas. I don't really know for sure. And why do I have this almost constant need to push myself to the limits? My mind must be going through some kind of transitional phase. I've got to try to figure it all out. I'm not depressed - not in the least. In fact, I'm very happy. It's just that I'm feeling weird. Almost out of sync.

March 8, 1992 (pg. 484) - At Palm Beach Community College, I will be speaking on April 9th about "How To Get Published".

April 7, 1992 (pg. 499) - I'm all ready for the Palm Beach and Lakeland speeches later this week.

April 9, 1992 (pg. 499) - I just got back from my talk up in Palm Beach. The speech went very well! I felt very confident and comfortable and I think it was my best talk ever!

April 12, 1992 (pg. 500) - The speech at the Sun 'n Fun EAA Fly-In up in Lakeland on Friday went well. Approximately 50 people showed up for my talk and slide presentation.

April 28, 1992 (pg. 513) - At BookFest this past weekend I got several business cards from people interested in "How To Get Published - Guaranteed". I met a woman who just self-published a book about cosmetic surgery. She also used Dan Poynter's "The Self-Publishing Manual" as her bible. Her format is a 6" x 9" softcover, $16.95, 160 pages. Her first edition is due out in July 1992. She's starting with 5,000 copies and is gearing up to sell them through beauty shops. Another woman came up to me and said that she heard me talk at last year's BookFest and was inspired by me and is now self-publishing a book. She is using Dan Poynter's book as a guide. On Sunday a guy came by to tell me he was already three-quarters of the way finished reading "Black Falcon". He had bought a copy on Saturday. He was really enjoying the book, especially since he was a pilot at Opa Locka Airport during the Second World War when it was a Naval Air Station for Dauntless dive bombers. Now he is a licensed delivery skipper and charter captain in the Bahamas and Florida Keys. He invited me to go sailing aboard a 45' sailboat that he operates out of Palm Beach. I have to get "How To Get Published - Guaranteed" written and published. I'm dragging on this project, but once again I've seen evidence that it will be a big seller for me.

May 8, 1992 (pg. 525) - "How To Get Published - Guaranteed" is seriously under way big-time. I hope to have it finished in six months.

May 12, 1992 (pg 528) - The woman who arranges the Sun 'n Fun EAA Fly-In speakers has invited me back for the 1993 air show.

May 17, 1992 (pg. 530) - I substituted at American High School 2 days this week and worked on "How To Get Published - Guaranteed" during that time, but it's been difficult to work on it at other times due to things popping up unexpectedly, like both of my vehicles failing emissions inspections, taking them to the gas station, then back for retesting. I need some solid time to keep the momentum going on this important book. I found a great passage in "The Autobiography Of Mark Twain". He wrote, "I proposed to be my own publisher," in reference to "The Adventures Of Huckleberry Finn". I finally found the proof I was looking for that he was indeed a self-publisher.

June 12, 1992 (pg. 554) - I finalized the chapter headings for "How To Get Published - Guaranteed" today. I've really been doing a lot of writing for "How To Get Published - Guaranteed". That's why I haven't been writing in this journal a whole lot recently.

June 18, 1992 (pg. 558) - Today I spent four hours going through all of my files and research material for "How To Get Published - Guaranteed" and organizing everything in file folders according to chapters.

July 13, 1992 (pg. 566) - I wrote a letter to the editor of "Aviation International News" last night to try to get him interested in doing a human interest story on me. I typed half of it this morning and will finish typing it tomorrow and have it in the mail by tomorrow afternoon.

August 3, 1992 (pg. 577) - In the mail today was an answer from the editor of "Aviation International News". He started off his reply by writing, "Your letter was one of the most interesting I've read in a long time." Then he went on to offer me two freelance writing assignments for his magazine. He offered to pay me $400 per article, with a 50% kill fee if the articles aren't used in the magazine. I'm going to take him up on the offer. I was suppose to start writing again this week on "How To Get Published - Guaranteed", but I just started writing the first story for "Aviation International News" instead.

August 5, 1992 (pg. 578) - The article for "Aviation International News" about corporate aviation is coming along strong.

August 9, 1992 (pg. 581) - I'm up to four pages on the article - very tightly written.

August 13, 1992 (pg. 582) - As of midnight tonight, I'm on vacation for nine days.

August 26, 1992 (pg. 582) - I never got a chance to catch up on this journal due to a busy vacation week, then we came home from vacation to find out a killer hurricane was heading towards South Florida, and then the storm (Hurricane Andrew) hit and we've been living for the past three days in the aftermath of the worst natural disaster this country has ever seen. So far $15 billion is the estimate to rebuild people's lives, homes and businesses. Homestead Air Force Base was leveled. Tamiami Airport was leveled with over 200 aircraft destroyed. All the trees in Cape Florida were flattened to the ground. Entire neighborhoods of houses were destroyed. Over 1.5 million people without electricity. All of Dade County under a 7pm to 7am curfew due to downed powerlines and looting. The National Guard has been called in to protect people and property (what is left of it). Over 100,000 people homeless due to destroyed houses. Hospitals destroyed. Shopping malls destroyed. Hotels destroyed. Flooding. Over 50,000 boats destroyed. Deaths. Water shortages. Ice shortages. Food shortages. The towns of Homestead and Florida City in South Dade were totally destroyed. The town of Kendall 90% destroyed. I could go on and on, but I won't.

August 28, 1992 (pg. 588) - I worked until 3:30am this morning unloading 4,000 pounds of supplies for storm victims. Then I got called in to work from 12:30pm until 5:30pm today (my day off) to help out with continuing hurricane relief flights. As of last night, U.S.

Customs Service Miami Air Branch has been using us as their temporary base because their facility at Homestead Air Force Base was wiped off the planet.

August 30, 1992 (pg. 589) - Things are getting worse in some ways and better in other ways with the hurricane relief efforts. Heavy rains last night and today caused even more damage and a lot of relief supplies got ruined (like clothes and bedding).

August 31, 1992 (pg. 589) - The damage estimate is up to $30 billion. 19,000 troops (Army, Navy, Air Force and Marines) have been ordered to assist South Dade County. Tent cities have been erected to house 2,500 people per tent. So far Broward County residents have donated 5 million pounds of food. Five U.S. Naval ships docked here with 2,000 tons of supplies. The Air Force has been flying in C-141 Starlifters and C-5A Galaxys about one every 15 minutes all day long. It's a full-scale military operation.

September 4, 1992 (pg. 590) - Not much time to write in this journal lately. Lots of overtime at work at the airport due to hurricane relief flights.

September 8, 1992 (pg. 603) - I'm trying to get things back to normal. I have to finish the article for "Aviation International News" and then start writing "How To Get Published - Guaranteed" again.

September 17, 1992 (pg. 611) - Today is the first day in over a month that I'll have a chance to work on the article for "Aviation International News". In fact, I'm going to do it right now. I've got four solid hours to work on it.

September 23, 1992 (pg. 613) - I was able to write a little more of the article for "Aviation International News" today. It's almost finished. Then I just have to polish it a bit and send it off. I want to get back to writing "How To Get Published - Guaranteed" so I can have it published for BookFest in April.

September 26, 1992 (pg. 616) - I finished the "Aviation International News" article yesterday. I'll type it on Monday, then send it to the magazine.

September 28, 1992 (pg. 616) - I went to type the article for "Aviation International News" today and realized I didn't have any plain white typing paper, so I had to go borrow some from a neighbor. I never got more than the first page of the article typed, so I'll have to finish it tomorrow and Wednesday.

September 30, 1992 (pg. 619) - I've got eight pages typed of the article. I'll try to finish it tomorrow. Looks like it will be almost fifteen pages typed, double-spaced.

October 1, 1992 (pg. 619) - Well, I said I was going to do it and I did it! I mailed off the article to "Aviation International News". Let's hope it gets published. It turned out to be twelve pages typed.

October 3, 1992 (pg. 619) - I started working on "How To Get Published - Guaranteed" yesterday while subbing. It feels great to be caught up and back on track with this book.

October 12, 1992 (pg. 622) - I've been writing like crazy on "How To Get Published - Guaranteed". I've completely finished Chapters One, Two and Three and am now going to type them. They came out longer than I expected, but they're very tight and informative.

October 14, 1992 (pg. 622) - I never got started typing Chapters One, Two and Three. Instead, I started writing Chapter Four. I should finish it by tomorrow.

October 20, 1992 (pg. 625) - I typed Chapter One today and organized my notes to begin writing Chapter Five.

October 22, 1992 (pg. 626) - I started typing Chapter Two of "How To Get Published - Guaranteed" today and will start writing Chapter Five tomorrow while I sub at school.

October 27, 1992 (pg. 627) - I finished writing Chapter Five yesterday and started organizing Chapter Six. Today I typed more of Chapter Two.

November 2, 1992 (pg. 632) - A lot has happened during the past four days. I wrote the entire Chapter Six on Friday and organized Chapter Seven. I also typed almost the entire Chapter Two. I got a letter of acceptance today from "Aviation International News" for my article! They'll be sending a $400 check within the next week or two. The article will appear in the November/December 1992 issue. It was shortened somewhat, but I don't know how much or what was edited out. I am truly excited about this article. It's going to be read by aviation professionals around the world!

November 4, 1992 (pg. 634) - I finished typing Chapter Two today, but still haven't had a chance to start writing Chapter Seven. Hopefully tomorrow I'll get around to it.

November 6, 1992 (pg. 634) - I started writing Chapter Seven while subbing today.

November 9, 1992 (pg. 637) - I finished writing Chapter Seven today and organized Chapter Eight.

November 11, 1992 (pg. 638) - I started writing Chapter Eight today.

November 14, 1992 (pg. 638) - My article came out in "Aviation International News" on Thursday. The article takes up two pages in the magazine.

November 16, 1992 (pg. 643) - I finished writing Chapter Eight today. I'll try to start Chapter Nine this week, but I've also got to get ready for this weekend's Miami Book Fair International, plus do some projects around the house. I also have to start typing more of the completed chapters. I took most of the photographs for "How To Get Published - Guaranteed" last week and I brought the roll of film in for developing today. I finished a rough mock-up of the cover. It looks good. The colors will be black and white on a bright red background.

November 17, 1992 (pg. 643) - I had to move up to the next larger size vault at the bank today. Writing sure creates a lot of paperwork. I organized the material for Chapter Nine today. I got everything ready for the Miami Book Fair International today.

November 18, 1992 (pg. 645) - I got the photographs for "How To Get Published - Guaranteed" today. They came out good for not having been taken by a professional photographer.

November 22, 1992 (pg. 645) - The Miami Book Fair International is over. I did very well and had a fun time, even though it rained on and off during most of the show. I was the most relaxed I've ever been for a show. Each show is getting easier. Throughout the entire three-day show I met a lot of interesting people. Everyone seems to have a story to tell - which is why "How To Get Published - Guaranteed" is going to be such a big success. I'm really psyched about "How To Get Published - Guaranteed".

November 25, 1992 (pg. 649) - I'm getting frustrated. I haven't been able to write or type "How To Get Published - Guaranteed" for almost the past two weeks due to projects like remounting the fence gate and fertilizing the lawn. I was on such a roll, completing a chapter a week for the period from mid-September through mid-November, then nothing. I haven't written anything since completing Chapter Eight. Now I've got Thanksgiving and the up-coming Christmas holidays, so I won't be able to write much until after New Years Eve.

November 30, 1992 (pg. 653) - I started writing Chapter Nine today.

December 1, 1992 (pg. 654) - I got my check from "Aviation International News" today.

December 3, 1992 (pg. 655) - Chapter Three is all typed. I'm still writing Chapter Nine. I hope to have it finished by tomorrow. It's been bits and pieces, here and there in writing Chapter Nine due to projects around the house.

December 5, 1992 (pg. 655) - I didn't finish writing Chapter Nine yesterday, but almost. I should be able to finish it Monday.

December 7, 1992 (pg. 658) - I finished writing Chapter Nine today.

December 8, 1992 (pg. 658) - Lots of yardwork. I'm trying to bring back the lawn and the fruit trees by pruning, seeding, fertilizing and watering.

December 13, 1992 (pg. 660) - I've been editing Chapter Two and it's grueling work because I'm basically rewriting the entire chapter. The next two weeks are going to be busy with family and friends.

December 17, 1992 (pg. 663) - I still haven't finished rewriting Chapter Two. It's the longest one in the book so far.

December 20, 1992 (pg. 664) - I finished rewriting Chapter Two. I'm going to try to retype Chapter Two tomorrow and organize myself for the writing of Chapter Ten.

January 4, 1993 (pg. 669) - I haven't written anything in almost two weeks. I'm way behind on "How To Get Published - Guaranteed" and there is no way that it's going to be ready by April - probably not until the summer now. In a way this is good because it takes a lot of pressure off of me.

January 11, 1993 (pg. 673) - I finally got a chance to organize Chapter Ten. I hope to start writing it tomorrow.

January 13, 1993 (pg. 674) - I started writing Chapter Ten yesterday.

January 21, 1993 (pg. 678) - The past two days have been full of killer yardwork again. I am in so much pain in my shoulders, it's unbelievable. I only have two more trees left to do - the key lime tree and the tangerine tree. I can't believe that I'm putting a potential bestseller on the back-burner for this @$#& [*4-letter word deleted*].

January 24, 1993 (pg. 682) - I finished writing Chapter Ten tonight. Tomorrow I'll organize Chapter Eleven.

January 29, 1993 (pg. 682) - I never got started on organizing Chapter Eleven until today due to the weekend with my family and more yardwork.

February 8, 1993 (pg. 686) - I finished writing Chapter Eleven today. I started typing Chapter Four this past weekend. I'm going to organize Chapter Twelve tonight.

February 9, 1993 (pg. 686) - I never got a chance to organize Chapter Twelve last night, but I'll do it today.

February 22, 1993 (pg. 695) - I started writing Chapter Twelve.

March 1, 1993 (pg. 697) - I wrote a bit of Chapter Twelve today.

March 3, 1993 (pg. 697) - I sent RFQs (Request For Quotations) to printers today to get rough estimates for printing and binding "How To Get Published - Guaranteed".

March 6, 1993 (pg. 698) - I wrote some more of Chapter Twelve in the high school library yesterday after subbing.

March 10, 1993 (pg. 704) - I finished writing Chapter Twelve today. I've changed the cover design for "How To Get Published - Guaranteed" slightly, but it's still not finalized.

March 13, 1993 (pg. 709) - I started writing Chapter Thirteen yesterday after organizing my notes and I'm half finished with it. My wife and daughter are napping right now, so I'm going to work on Chapter Thirteen a bit more.

March 15, 1993 (pg. 709) - I finished writing Chapter Thirteen today.

March 17, 1993 (pg. 710) - I'm all organized to begin Chapter Fourteen.

March 18, 1993 (pg. 711) - I sent in the Advanced Book Information form on "How To Get Published - Guaranteed" for the "Books In Print" reference guide. I'm going to try to start writing Chapter Fourteen tomorrow while subbing. Next week I'll send for a Library of Congress Catalog Card Number. I've already picked an ISBN Number for "How To Get Published - Guaranteed".

March 19, 1993 (pg. 711) - I started writing Chapter Fourteen today.

March 29, 1993 (pg. 716) - I took some preliminary cover photos for "How To Get Published - Guaranteed". The new mock-up is almost done. I just have to do the border stripe and glue on the photo when I get it back from the store.

March 31, 1993 (pg. 716) - I finished the new cover mock-up. It looks great! It looks professional. I wrote more of Chapter Fourteen yesterday and a bit more today.

April 5, 1993 (pg. 718) - On Friday I got a phone message from the editor of "The Southern Aviator" magazine. His publication also prints the daily "Sun 'n Fun Today" newspaper at the air show. I called him and he wants to write a feature article about me for the April 20th issue of "Sun 'n Fun Today" (the day of my talk at the Sun 'n Fun EAA Fly-In), plus use excerpts from "Black Falcon", plus use a black and white photo of me - so I sent him the necessary information and excerpts. The Sun 'n Fun EAA Fly-In is really going all out to publicize my talk this year.

April 8, 1993 (pg. 718) - So far I haven't had much time in April to write in this journal because I've been so busy working on "How To Get Published - Guaranteed".

April 27, 1993 (pg. 726) - What a week!!! First Sun 'n Fun, then BookFest. So much has happened that needs to be written about, plus I'm so far behind in everything else that I'm still trying to catch up. The highlight is that I finally got to fly a Piper Tri-Pacer. I left for Sun 'n Fun at 7:30am last Tuesday and got to the air show by 11:30am. I did an okay job at my talk, but I was nervous as hell in front of around 200 to 300 people. The next day (Wednesday) was when I got to fly the Tri-Pacer. I got home late Wednesday night, and on Thursday I had to unpack from Sun 'n Fun, pack for BookFest, then go to work at the airport. On Friday I drove up to Palm Beach to set up my display for BookFest. It was the best display I've had yet. I spoke to a lot of people about "How To Get Published - Guaranteed" and had a lot of fun.

May 6, 1993 (pg. 742) - I am so close to finishing Chapter Fourteen. Then it's just a matter of wrapping up the book in Chapter Fifteen.

May 13, 1993 (pg. 744) - I finished Chapter Fourteen today, so I'm basically done with the book, except for the closing chapter titled "The Rewards Of Being A Published Author".

May 27, 1993 (pg. 746) - I'm just about finished with the closing chapter. Maybe I've got three or four pages at the most, which I hope to write tonight. If I can't finish it tonight, I'll try to finish it tomorrow afternoon.

May 30, 1993 (pg. 746) - It is now 3:38pm and I just finished the first draft of "How To Get Published - Guaranteed".

June 3, 1993 (pg. 751) - I haven't had time to do any typing this week. I'm subbing right now, so as soon as I finish writing in this journal, I'm going to organize all of the photos and figures for each chapter of "How To Get Published - Guaranteed".

June 10, 1993 (pg. 753) - I'd like to have "How To Get Published - Guaranteed" completely typed by the time I go on vacation in July. This week I finally got all the photos and figures organized. I made photocopies of them and put the masters in the vault, so I'm covered in case of disaster.

June 24, 1993 (pg. 761) - Lots of typing getting done. I'll be finished typing Chapter Eight today.

July 1, 1993 (pg. 763) - I haven't been able to type "How To Get Published - Guaranteed" in almost a week.

July 7, 1993 (pg. 764) - I haven't had a chance to type any more of "How To Get Published - Guaranteed".

July 14, 1993 (pg. 766) - I designed the page layouts for "How To Get Published - Guaranteed" yesterday.

July 15, 1993 (pg. 767) - Yesterday I was finally able to start typing again, but just barely. I'll try to get in some more today, but I'm going to be busy for the next week preparing for a vacation.

July 20, 1993 (pg. 770) - I finished typing Chapter Nine.

July 21, 1993 (pg. 772) - Well, I'm putting this journal and all documents and writing associated with "How To Get Published - Guaranteed" into the vault this morning for safe-keeping during my month-long vacation, so I won't be making any entries in here for a while.

August 19, 1993 (pg. 772) - I'm back from my trip, but I'm still on vacation from work for three-and-a-half more days. So much happened during the trip - new places, new people, new experiences. I took notes, so I'll rewrite the notes into this journal to give an overview of what happened. I still haven't figured out how many miles we traveled, but I'd estimate it to be around 10,000 air miles and 3,000 car miles, plus a lot of hiking - so about half the distance around the world in total.

August 20, 1993 (pg. 786) - It's back to the old grind again and I'm not happy about it. I mowed the lawn yesterday. It's back to work at the airport tonight after a month off. This vacation has been the longest period I've ever had off from school or work since I was 12 years old. I got used to not working. I'll try to get started on "How To Get Published - Guaranteed" this week.

August 31, 1993 (pg. 790) - I finally started typing "How To Get Published - Guaranteed" today. Also very busy designing the front and back cover copy, as well as all of the front and back matter (title page, copyright page, footnotes, etc.).

September 1, 1993 (pg. 790) - I'm really trying to take "How To Get Published - Guaranteed" big-time. I just sent for a Bookland EAN barcode for the back cover, which is almost a requirement for all books to survive in the world of large-scale publishing. Barcodes are the computerized scan bars that appear on most products these days. I didn't use the barcode on "Black Falcon" because I was afraid of computer technology.

September 3, 1993 (pg. 791) - I finished typing Chapter Ten today. I'll type Chapter Eleven tomorrow.

September 7, 1993 (pg. 791) - I'm working on the front and back matter for "How To Get Published - Guaranteed" some more. I'll try to start typing Chapter Eleven tomorrow.

September 15, 1993 (pg. 794) - I've been working like crazy on the typing of "How To Get Published - Guaranteed". I finished typing Chapter Eleven and started typing Chapter Twelve.

September 17, 1993 (pg. 797) - I finally finished writing most of the front and back matter for "How To Get Published - Guaranteed" and also redesigned the front and back covers and rewrote all the cover copy. That's a lot for a 6-hour day.

September 22, 1993 (pg. 798) - I finished typing Chapter Twelve today.

September 25, 1993 (pg. 799) - I got a call last night from a woman who heard me talk up in Palm Beach about a year-and-a-half ago. She has just self-published her novel because of my talk.

October 1, 1993 (pg. 802) - Once again, I haven't typed in about a week. I've been so busy getting my catalog ready for the upcoming Miami Book Fair International and writing updated RFQs for the printing and binding of "How To Get Published - Guaranteed". I still don't know which of the four book printers I'll be using. They're all good companies. Hopefully, I'll start getting quotations from them next week so that I can finalize my decision.

October 5, 1993 (pg. 802) - I will start typing Chapter Thirteen today as soon as I finish eating lunch.

October 11, 1993 (pg. 808) - I finished typing Chapter Thirteen today.

October 13, 1993 (pg. 810) - I've received 2 of the 4 quotes from printers, with a letter from another of the 4 saying that a quote is being prepared.

October 15, 1993 (pg. 813) - I am finally at the saturation point of frustration over my book! Next week I've got to go to work early at the airport on three out of five days to cover for people on vacation. And the holiday seasons are coming up (Thanksgiving, Christmas, New Years), which means a lot of relatives will be down on vacation, which means putting the book on hold for a while. Last year during the holiday season, "How To Get Published - Guaranteed" was put on hold for almost a month. I hope I don't have to do that this year, especially since I'm so close to finishing. I now fully understand why authors go to writing colonies. No phones, no yardwork, no job, no people to deal with, nothing to do except write, eat and sleep. Everything sure is catching up to me. I've got the Miami Book Fair International to think about. It's only a month away and I still haven't finished my new expanded display boards. I've got the typesetter to think about. He's expecting my manuscript by the end of October, but now that's just not possible. Hopefully, I'll have it finished by the end of November. The typing of the manuscript just keeps getting pushed aside because of things that come up in my everyday life. I wanted to type Chapter Fourteen this week, but couldn't even get started on it. Then I've got Chapter Fifteen, which I've decided to move up to Chapter Four, thereby making the previous Chapter Four into Chapter Five, and Chapter Five into Chapter Six, and so on. Then I've got the new final closing chapter to type, which is Chapter Sixteen. All this time that I've been typing the manuscript so far, I've also been editing as I've gone along, so that the typed version is really a second draft of "How To Get Published - Guaranteed". After I finish all the typing, I'd like to go over the manuscript one more time and do a final editing, which will be the third draft. I am so far behind where I planned to be at this point that the only way I can possibly try to catch up is to push my personal life out of the way for the next month or two.

October 18, 1993 (pg. 814) - I got another printing and binding quotation. So far McNaughton & Gunn is the 1st choice. I finally got around to start the typing for Chapter Fourteen. It's a long chapter and will end up being Chapter Fifteen in the book once I change all the chapter numbers.

October 19, 1993 (pg. 814) - I've got all the printing and binding quotes. I chose McNaughton & Gunn as the printer/binder for "How To Get Published - Guaranteed".

October 21, 1993 (pg. 815) - I've been working on and off in little spurts all week on the typing of Chapter Fourteen, but so much other stuff keeps popping up that it's been difficult to get much progress done on it. I'm only about one third of the way through with it.

October 23, 1993 (pg. 816) - Yesterday, while subbing, I rewrote Chapter Sixteen, the new final chapter of my book. It's a much better ending now.

October 28, 1993 (pg. 824) - I'm almost finished typing Chapter Fourteen, which is really Chapter Fifteen.

November 1, 1993 (pg. 825) - Today I finished typing Chapter Fourteen and also typed Chapter Sixteen. I skipped the old Chapter Fifteen because I've got to change it to Chapter Four and move all the rest back one.

November 9, 1993 (pg. 828) - My wife and daughter are flying up north tomorrow for 5 days. I'll spend that time finishing my book display.

November 13, 1993 (pg. 831) - The past two-and-a-half days have been spent on the book display. The display is finished. It looks great. Friday, Saturday and Sunday is the Miami Book Fair International. I hope it all goes well.

November 15, 1993 (pg. 832) - I'm all set for the Miami Book Fair International, so I'll spend this week typing Chapter Fifteen, which is going to be Chapter Four.

November 17, 1993 (pg. 833) - I never got around to typing Chapter Fifteen yesterday due to tons of errands that had to be done (problems with the cable television, problems with the air-conditioning, problems with a bathroom sink drain), but I was able to type about half of it today and will finish it tomorrow. Now that I've changed this chapter from Fifteen to Four, I'll refer to it as Chapter Four from this point forward. I'm getting close to the production stage of "How To Get Published - Guaranteed". Two more days until the big book show!

November 18, 1993 (pg. 834) - I didn't have much time to type today. I have around 3 or 4 more pages left to go before the text is finished.

November 22, 1993 (pg. 835) - YES! YES! YES! This year's Miami Book Fair International was a huge success - my best event ever since I started publishing my books. I finally hit on a winning combination of books. I also met many people who want me to be a guest speaker. If all these come to fruition, I will be speaking a lot this coming year. I'll try to finish up typing Chapter Four this week, but I don't know if I'll have time due to preparations for a big Thanksgiving party we're having at our house.

November 24, 1993 (pg. 836) - I forgot to mention that at the Miami Book Fair International I met the owner of a company that owns DC-3s, DC-6s and other radial-engined airfreighters. He gave me his name and phone number and said that he could arrange for me to fly jumpseat in the cockpit aboard his planes.

November 30, 1993 (pg. 840) - I'll finally be doing some typing tomorrow after nearly 3 weeks of no typing. Hopefully I'll finish up Chapter Four.

December 1, 1993 (pg. 841) - I finally finished typing Chapter Four, which means "How To Get Published - Guaranteed" is all typed. Now I just have to do a final editing, then send it off to the typesetter. One of the changes I've decided to make is to change Chapter Four, which used to be Chapter Fifteen, into Chapter Two. Confusing? Maybe, but hey, that's life! It sure feels good to finally be finished with "How To Get Published - Guaranteed" and to have the typesetting and book production just around the corner. It's been 3 years since the idea for the book was originally conceived and 21/2 years since I started working on it.

December 9, 1993 (pg. 844) - I'm speeding along so far on the final editing.

December 15, 1993 (pg. 845) - Yesterday I saw a banner plane towing a banner that said, "Live To Fly, Fly To Live." How True!

December 17, 1993 (pg. 846) - Starting this weekend is when all the relatives start coming down. It's going to be a long, busy holiday season.

December 19, 1993 (pg. 846) - Yesterday I discovered that I had left my electronic typewriter turned on for 2 days. I was worried that the circuits may have burnt out, but it seems to be working fine.

December 24, 1993 (pg. 848) - Slowly progressing on the final editing - at a snail's pace!

December 27, 1993 (pg. 849) - Slow to none on the progress of the final editing.

December 29, 1993 (pg. 849) - I finished the final editing of Chapter Five, so I'm one third done.

December 31, 1993 (pg. 849) - All the relatives go back up north in a couple of days. The end of another year. The beginning of a new one. This is The Year that I will become a 2-book author. No longer will I be a 1-book author. This is good. No! This is fantastic!

January 2, 1994 (pg. 850) - I had to go in to work early today for a killer day - lots of jets.

January 6, 1994 (pg. 853) - Yesterday I was reading an aviation classified magazine called "Aero Trader". I found a Florida Keys home on a private 2,400 foot grass airstrip, with dockage, an ocean-access canal, and a marina down the block. PARADISE! Someday! Hopefully!

January 12, 1994 (pg.856) - I got a call today from the Forums Chairman for the Sun 'n Fun EAA Fly-In. Someone canceled for Tuesday of the show, so he's giving me that time slot. Now I'll be giving 2 forums because the original time slot I had for Thursday is still mine. I'm finally making progress with the final editing of "How To Get Published - Guaranteed". Over half done!

January 13, 1994 (pg. 857) - I'm up to Chapter Twelve on the final editing. This week I'd like to finish Chapters Twelve and Thirteen, then next week do Chapters Fourteen, Fifteen and Sixteen, then send the text to the typesetter while I type the front and back matter.

January 18, 1994 (pg. 861) - I finished Chapter Twelve's final editing yesterday.

January 21, 1994 (pg. 862) - The past 3 days have been a marathon of little or no sleep. I got up at 7:00am on the 19th, spent the day doing errands with Bonnie, worked at the airport until 2:30am on the 20th, got to bed by 3:30am on the 20th, woke up at 6:30am on the 20th to get Jen ready for school, spent the day doing errands and editing a couple of pages from Chapter Thirteen, worked at the airport until 1:15am on the 21st, got to bed by 2:15am on the 21st, woke up at 6:00am to sub at American High School, finished editing Chapter Thirteen while subbing, took Jen to the playground, took Jen out to dinner while Bon was out with her friends, and now it's 8:30pm on the 21st. I'll be up until at least midnight typing some of the edited Chapter Thirteen. In total, it will be 7 hours of sleep during a 67 hour period. That's 60 hours AWAKE during that same 67 hour period. I'm wired!

January 22, 1994 (pg. 863) - I ended up staying awake until 2:30am today to type most of the edited Chapter Thirteen, so I was awake for approximately 63 hours during a 70 hour period. Today I'll try to finish typing the edited Chapter Thirteen and begin the final editing of Chapter Fourteen.

January 23, 1994 (pg. 863) - I'm on a roll with "How To Get Published - Guaranteed". I edited almost all of Chapter Fourteen yesterday and stayed up until 3:30am this morning typing the edited Chapter Thirteen. Bon thinks I'm going to burn myself out, but I've got to do it to have my books by the Sun 'n Fun EAA Fly-In in April.

January 26, 1994 (pg. 864) - I finished typing the edited Chapter Fourteen, finished editing Chapter Sixteen (skipping over Chapter Fifteen, which is the longest chapter in the book), started editing Chapter Fifteen, and will try to type the edited Chapter Sixteen tonight. I won't be done with Chapter Fifteen this week because I've been working a lot of overtime at the airport, and because Bon, Jen and I are going down to the Keys on Friday. So, hopefully, early next week! In fact, I'm canceling flying next Thursday to finish the book. That's rare!

January 29, 1994 (pg. 864) - Had a blast in the Keys yesterday. Spent the day in Islamorada swimming, snorkeling, playing in the sand with Jen and enjoying the scenery in Florida Bay. It reminded me why I've been working so hard on "How To Get Published - Guaranteed", which is to be able to live in a place like the Keys! Bon and Jen are napping right now, so I'm going to work on editing Chapter Fifteen. The edited Chapter Sixteen is all typed.

February 3, 1994 (pg. 867) - YES! Today's the big day. I finally finished the final version of "How To Get Published - Guaranteed". It's going to the typesetter the first of next week. I spoke to the V.P. of the typesetting company this morning. Everything is set! I don't think it has really hit me yet (emotionally) that I've just completed my 2nd book. It's been so long a time since I first started jotting notes for it.

February 9, 1994 (pg. 869) - I finished reading "How To Get Published - Guaranteed" from start to finish for the first time. It's a good book, but I made it better by doing one more final rewrite (the 4th draft). I made about a dozen major changes (entire paragraphs added, deleted or severely changed) and about 400 minor changes (one or two words changed). This all took me about 4 days. Today I typed all the corrections. On Friday I'll do the final paste-up of the corrections, make the photocopies (if I have time), and ship it to the typesetter (if I have time).

February 13, 1994 (pg. 870) - I never got to do the paste-up of the final corrections for the 4th draft due to lots of errands in my personal life and lots of overtime at the airport. Delays, delays and more delays. Do I know what frustration means? Most definitely.

February 16, 1994 (pg. 873) - I'll have a chance to start the paste-up tomorrow night. Then I'll finish this weekend while Bon and Jen are out of town, which means that if no other delays pop up, the manuscript will be sent to the typesetter for sure on Monday morning.

February 19, 1994 (pg. 874) - I pasted up about a third of "How To Get Published - Guaranteed" the other night, but Bon and Jen did not go out of town this weekend, so I spent last night (11:00pm to 2:30am) pasting up another third of the book, did a bit more this afternoon while Bon watched her soap operas, and will finish it tonight after Bon goes to bed. Then I will white-out all the pen markings.

February 21, 1994 (pg. 874) - I stayed up until 2:30am again the night before last and finished the paste-up and most of the white-out of pen marks. I finished the white-out last night and also started the final reading of the manuscript. It's finally perfect!

February 22, 1994 (pg. 874) - I was too exhausted to stay up late last night, so the final reading never got finished. I thought I was through with delays, but they just keep coming up. I'm feeling very pressured right now.

February 23, 1994 (pg. 875) - I finished the final reading of "How To Get Published - Guaranteed" this morning, but I got called in to work early at the airport for overtime, so I couldn't photocopy the final draft. I will definitely do it first thing tomorrow morning. I'm changing the publication date to October 1994 (with advanced copies available in May 1994), which will take the pressure off of me and give me a good six months to do advanced publicity, press releases, etc. It's better to take my time and do things right than to rush it. After I send off the manuscript tomorrow, the next time I read "How To Get Published - Guaranteed" it will be in page format. I'm excited. It's been a long haul!

February 25, 1994 (pg. 875) - I photocopied the final draft yesterday and wrote detailed instructions for the typesetter, but I didn't have time to ship everything. I will definitely do it today because there's nothing else to do to the manuscript (except get it typeset). I'm very pleased with the results. It was worth the extra effort I put into the final draft.

February 26, 1994 (pg. 875) - Well, I shipped the manuscript to the typesetter yesterday.

March 1, 1994 (pg. 878) - My typesetter called yesterday to say the manuscript arrived.

March 28, 1994 (pg. 901) - I got the cover mechanical proof today. Excellent! Only a couple minor changes, which should be done tomorrow.

March 29, 1994 (pg. 902) - I spoke to the typesetter who's doing "How To Get Published - Guaranteed". The text is finished and was sent UPS yesterday, so I should get it tomorrow. The front and back matter will be ready to ship on April 11th, but I'll be at the Sun 'n Fun EAA Fly-In, so I'm having them ship it on the 15th of April to make sure I'm home when it arrives. I don't want to take a chance of it sitting on my front porch and getting caught in a rainshower. UPS is always leaving things unattended on the porch. I checked out the computer set-up that Bon picked out. It goes on sale April 1st, which is this Friday. We'll buy it first thing Friday morning. I'm hoping to learn enough by the beginning of June to do my promotional work for "How To Get Published - Guaranteed" on it.

March 30, 1994 (pg. 903) - The page proofs for "How To Get Published - Guaranteed" just arrived. Unbelievable! It's a book! FINALLY! After 31/2 years! I'm going to Office Depot right now to make photocopies of them. The text is 270 pages, so my book will be well over 300 pages. I underestimated the page count a lot! But, heh! Imagine that! A 300+ page book! I AM SO EXCITED!

April 3, 1994 (pg. 904) - Bon and I bought a computer on Friday. We got a Packard Bell with a Hewlett Packard ink-jet printer. I spent all day and all night yesterday going over my page proofs with my dad. He read the manuscript and I followed along on the page proofs. Only 2 chapters left to do, which we'll do this morning. Only 1 major typesetting mistake so far, where 3 lines of manuscript were left out. But it's not so major because 4 blank lines were left on that particular page. There are several dozen typos, but nothing major - NOTHING as bad as the typesetting job that was done for "Black Falcon". I'm very pleased.

April 6, 1994 (pg. 906) - I got confirmation from my printer that the increased page count and the 4-color process for the cover will only add about $700 - $800 to the total cost.

April 27, 1994 (pg. 921) - I finished editing the page proofs for the front and back matter today.

April 29, 1994 (pg. 922) - Bon and I are buying new office furniture for Tri-Pacer Press because the computer takes up my entire desk, leaving me with no place to work. Now I'll have a real professional set-up with cabinets, computer, large desk top, good office chair, etc.

May 2, 1994 (pg. 923) - I spoke to the typesetter today. I could possibly have the actual camera-ready copy by next Friday. So I'm probably on the final month prior to actual book production.

May 14, 1994 (pg. 936) - Had my first computer class today. It was good, but the first 2 hours went very slow. Next Saturday we'll actually run programs. Today was just tutorials.

May 18, 1994 (pg. 937) - Orders for "How To Get Published - Guaranteed" are trickling in via mail-order and telephone calls from people who have heard about the book. The first confirmed bookstore order came in from a California-based independent bookstore. They had sent a check a couple of weeks ago, but I didn't deposit it until I could get confirmation that they want the book on back-order. I got the confirmation in the mail yesterday.

May 21, 1994 (pg. 938) - I finished the photo boards yesterday, the cover mechanical overlay was finished the day before and the camera-ready copy arrived yesterday - so all that is left is to paste up the figures, then make photocopies, then ship all the supplies to the printer - hopefully by Friday. The second computer class was excellent today. Learned a lot of neat stuff to do on the computer in regards to word-processing, editing, illustrating, data bases, etc. I can't wait until I send off the book stuff to the printer so that I have time to play on the computer all day and start doing the press releases!

May 23, 1994 (pg. 939) - I used the printer on the computer for the first time today. Little by little I'm learning new things. The start of the paste-up process was frustrating today. The lightboard did not work due to the bulbs being too low wattage.

May 24, 1994 (pg. 939) - I finally figured out how to make the homemade lightboard work. Paste-up went smooth today.

May 26, 1994 (pg. 941) - I finished the paste-up last night after work, staying up until 2:00am this morning. I spent today finalizing my instructions to the printer and making photocopies of the text, the photos, the mechanical and the mechanical overlay. I will package up everything tonight after work. Tomorrow I will go to UPS and ship it all to McNaughton & Gunn up in Michigan. Then the book will be out of my hands until the bluelines arrive.

May 28, 1994 (pg. 942) - The stuff for "How To Get Published - Guaranteed" was shipped out at around 2:45pm yesterday on Next Day Air UPS. I called McNaughton & Gunn and they'll call me as soon as it arrives.

June 6, 1994 (pg. 944) - I got an inquiry in the mail from Alaska in regards to "How To Get Published - Guaranteed". I sent the woman an order form.

June 14, 1994 (pg. 947) - I'm having a blast with the computer. I really WAS working in the "dark ages" previous to learning how to use it. I still have a lot to learn, but at least I know enough to do my press releases and publicity work this summer.

June 21, 1994 (pg. 948) - I received the text blueline proofs yesterday. Only a few very minor mistakes, all of which are easily correctable. I shipped the proofs back to the printer today. Tomorrow the cover proof will arrive. Then the actual printing will begin.

June 22, 1994 (pg. 949) - The cover proof came today. Impressive! Only a couple of very minor blemishes that need to be corrected. I'm shipping the proof back right now for the corrections to be done.

July 18, 1994 (pg. 958) - I just got a finished, bound copy of "How To Get Published - Guaranteed". Perfect! The book is in print.

Now you know the entire story behind the writing and publishing of *How To Get Published - Guaranteed.* It took about three-and-a-half years to take this project from an initial idea to a bound book, with many interruptions in between, such as Hurricane Andrew, overtime at my job, vacations, speaking engagements, books shows, trade shows, and lots of yardwork. If *How To Get Published - Guaranteed* had been written on a full-time basis, I probably could have completed it in less than a year. Unfortunately, that option was not available to me. By doing it on a part-time basis, however, I was able to prove once again that an ordinary person with enough drive and determination can successfully write and self-publish a book, even while leading a very busy lifestyle. *How To Get Published - Guaranteed* went on to win a book award from the National Association of Independent Publishers.

It should be obvious to you after reading *Birth Of A Book* that my personal journal has become a major part of my life, as well as a major undertaking. I write in it a lot - sometimes for several hours per day. What do I write about? Everything! To give you a better understanding about who I am as a person and what I write about, the following journal entries have been selected for your reading enjoyment. I have included these daily entries in their entirety, so each one is much more lengthy than the passages I excerpted for *Birth Of A Book.*

June 16, 1990 (pgs. 23 - 26) - Bon, Jen and I have been through so much in such a short time. Together, the three of us have seen how close life and death are to each other. Although Jen is only 21/2 years old, for a while we thought we were going to lose her because of a heart defect. She has been through open heart surgery (plus 2 other surgical procedures) and has pulled through in an excellent and strong way and her cardiologist says she is doing fine and should now live a normal life expectancy. But before she had that surgery, we had to face the following dilemma: If she didn't have the surgery, she would only have lived to be about 6 years old, which means that her life would now be almost half over. The other part of the decision was that when she went in for surgery, we could have lost her during the operation due to possible complications. This really sunk in at two specific times. One was when we had to sign our names to a card stating that we understood the risks involved and that Jennifer (I've always loved that name and knew that if I had a daughter I would name her Jennifer) could die from the surgery and that we would not hold the hospital responsible. The second time was when another couple we knew, whose child also had a heart problem, put their child in for open heart surgery a week before Jennifer's operation and their child didn't make it. I'm crying for them right now for the experiences they have missed, for not being able to watch their child grow up. I know we made the right choice, though, because Jennifer sat up in her hospital bed after only 3 days in ICU and she came home after only 6 days total and she is a beautiful, active little girl. And she knows all about her scar, which is about 5 inches long down the middle of her chest. She likes to show me her scar and then she points to a scar on my right leg and we both laugh while comparing our scars. Jennifer is so much fun. She loves to look at sailboats with me and come to the airport with me to watch airplanes taking off and landing. And every time she spots a plane in the sky she points to it and says, "Daddy, plane!" and she gets so excited. I can't wait until she's old enough to know not to grab the controls so that I can take her flying with me. She's already been flying all over the country aboard jetliners and she's also been on a short flight out of North Perry Airport aboard an antique airliner, a 1929 Ford Tri-Motor. She loved it. I took her to an air show a couple of months ago. At the air show I brought Jennifer into the cockpit of a Douglas Super Three and she sat in the captain's seat while I sat in the copilot's seat. I showed her all the controls and she played with the yoke and then she stuck her little head out the opened side cockpit window and waved to everyone down below on the ramp. And then later we sat beneath a Navy F-16 fighter jet and ate lunch and we both had a great day together. Jennifer ate her first hot dog at that air show. She is so much her daddy's little girl and so much alive and yet she could have been so close to death only a year earlier when she went in for the open heart surgery. I can remember when the orderlies were wheeling Jennifer from her hospital room down to the operating room and I watched the stretcher go around the final turn until I couldn't follow them anymore and thinking that I might not see Jennifer alive again. And then a few hours later I was allowed to see her after the surgery and she was lying on her bed in the Critical Care Unit, naked, on her back, her arms and legs tied down on the sides of the bed so that she wouldn't kick or pull out her tubes and wires, which were many and going to all parts of her little body. She was hooked up to several machines and also had a pace maker going to her heart temporarily and monitor wires and a chest drain and several IVs in her arms and legs and a breathing tube down her throat and another tube going into each nostril and a cathoder, and as I was looking at my little girl I was thinking, "This isn't something a person is suppose to see their child going through. This is what a person expects to see their parents going through in their old age." I grew up a lot during those 6 days Jennifer was in the hospital. And I discovered something. Jennifer's problem was minor compared to what other kids were in for. When Jennifer was switched on the 3rd day from the Critical Care Unit to the Step-Down Ward, the kid in the bed next to her had been paralyzed since birth and had spent most of his life in a hospital bed. And his parents were there, still spending time with him. What had they been thinking during that time, not being able to take their son to air shows or the beach or the grocery store? And what were they thinking about the future? Was it a living death for them? It hurts when I see parents standing by their child in that kind of situation, and then I see parents of normal, healthy children who abuse their kids, both verbally and physically just because of some minor incident the kid may have caused. It amazes me how a parent can abuse their own child.

November 22, 1990 (pg. 116) - I bought Jennifer her first model airplane. It was a standard $1.99 balsa wood, rubberband-powered flyer. We went out front today to fly it. It took several trys to get it into proper trim so that it wouldn't bank to the left. But then it would only climb and stall. Then Jennifer accidentally broke off a piece of the delicate horizontal stabilizer, so I figured that we'd fly the plane until it crashed and really have a wild time with it. So I wound the rubberband extra tight for more power and instead of throwing the plane horizontally, I threw it vertically. It was the best flight of the afternoon, climbing straight up to about 75 feet, then did all kinds of aerobatics on its own, then crashed into the grass. But it didn't break apart. Then Bonnie came outside to watch, so we tried to show her the same thing, only this time the airplane landed in a Black Olive tree in our front yard. It's still stuck up there - about 40 feet above the ground. Oh, well. We'll just have to buy another one. It sure was fun.

December 21, 1990 (pgs. 132 - 135) - Bonnie and I went out on the sailboat today. The weather forecast was calling for sunny skies all day with 15 knots of wind out of the east. The actual winds were up around 20 knots. Very breezy out on Biscayne Bay. Lots of chop. We launched from the dock on Key Biscayne by 10:00am, motored out the channel and were under sail by 10:30am. "Dreamer" has a roller reefing system for the mainsail, which I've never had to use prior to today. Because of this, I had to dig into my spare parts bin to find a 1/4" bolt for the reefing system lock, which was missing. I had the right size bolt. Bonnie and I reefed down the mainsail to the equivalent of 2 reefs in a standard reefing system. This shortened the sail enough to give us good speed, but with little heel. We raised the working jib for part of the time, but took it down for awhile, and then hoisted it back up. I didn't have a speed log, but I estimated that we were doing between 3 knots and 6 knots at various times throughout the sail. 6 knots is the designed hull speed for "Dreamer", which is really moving for this type of boat. In total, we sailed from 10:30am until 1:00pm for a distance of about 9.5 nautical miles. This gave us an average speed of about 3.8 knots. Not bad for an old twenty-two foot trailer-sailer with worn-out sails. Our course took us from Key Biscayne across the bay on a starboard tack to Coral Gables Estates, where we checked out all the waterfront mansions. We then headed on a northerly course, sailing the opposite tack past Coconut Grove and Dinner Key, up to Mercy Hospital, then with a controlled jibe we headed out into the bay for about a mile and a half before heading north again on the port tack toward the channel beneath the Rickenbacker Causeway. We dropped the sails just short of the bridge and motored underneath to search for a quiet place to anchor for a couple of hours to read and rest. By this time, the clouds had rolled in and it got chilly. This was not forecasted. A light rain fell for a few minutes, but then eased up, only to come and go for the next two hours. We had planned on motoring up to Miami's Marine Stadium to drop anchor (like we've done many times in the past), but they've now placed 2 large "No Trespassing - Police Ordinance" signs in the middle of the cove to keep boats out. So, in defiance, we anchored just shy of the line between the 2 signs, over towards the north side of the cove by the Causarina pine trees. It was very peaceful there. We were the only boat in the cove. Today's sail is what the sport of sailing is all about - a brisk cruise at near hull speed under blue skies, followed by a period of total relaxation and solitude in a quiet, tree-lined anchorage - and damned the signs and the rain. It felt fantastic to be out on the water and in control of a sailboat. As an aside, an old Boeing Stratocruiser airfreighter flew overhead my truck today at approximately 1500 feet while I was towing "Dreamer" on the Rickenbacker Causeway toward the boat ramp. This was a rare aviation sighting (since only a few of these aircraft are still airworthy in the entire world) and it was the first time I've ever seen a Stratocruiser in flight, although I've seen half-a-dozen at various airports and air museums around the country. These aircraft used to be the queens of the skies at such airlines as Pan American during the early 1950s. Fortunately for me, the airspace around Miami International Airport is still a haven for lovers of old radial-engined transports (of which I am a true devotee). Just today, for example, between 9:30am when I was towing "Dreamer" on the Rickenbacker Causeway and 4:30pm when I was motoring back to the dock, I spotted the Stratocruiser, a heavily-laden Douglas DC-6 airfreighter still at only about 500 feet above the water heading in the direction of the Bahamas, a Douglas DC-3 off in the distance, another DC-6 which passed at about 1000 feet directly above "Dreamer" while I was sailing on the bay, a third DC-6 headed off in the distance for points south while I was anchored in the cove and a fourth DC-6 airfreighter which flew overhead while I was motoring back to the dock. Of course, in that same time period several dozen jetliners passed through the Miami TCA, as well as an Air Force Lockheed C-130 cargo transport and many corporate jets, but in my mind these aircraft can't compare to the old avgas-burning transports from an aviation era gone by. I'm lucky enough to live in an area where these aircraft still etch out a living flying goods to and from the Bahamas, the Caribbean, and South and Central America. Some of these aircraft are over fifty years old - older than the fathers of some of the pilots flying them. So far I've flown aboard a Douglas DC-3, a Curtiss C-46, a Beech-18, and a Ford Tri-Motor. Hopefully I'll add to this list in the future. I'd love to ride in the jumpseat of a DC-4, the main aircraft in my novel. I got to sit in the cockpit of the C-46 and the Tri-Motor, both thrilling flights. The C-46 was a cargo flight from Miami to Nassau and back. The Tri-Motor was a local flight out of North Perry Airport. It is such a pleasure to mix flying with sailing, and vice-versa. Just two days ago I enjoyed a fantastic flight up to Jupiter and got to fly above the hundreds of sailboats anchored in the Intracoastal Waterway and in the marinas from Fort Lauderdale to Jupiter, and then overhead dozens of sailboats that were out sailing on the ocean. And today I got to enjoy watching airplanes while I was the one doing the sailing out on Biscayne Bay. And now I'm mixing my memories of flying and sailing with writing. To end the day, I'm going to make love with my wife. How can this be topped? It can't!

December 26, 1990 (pg. 145) - Why is it that when I land after a flight the feeling is not the same as after I've mowed the lawn or cooked a steak on the BBQ or fueled a corporate jet? Not much can compare with the feeling of having just flown an aircraft.

February 27, 1991 (pg. 207) - When I don't write an entry in this journal for several days, it doesn't mean that I don't have anything to write. It means that either I was out of town, didn't have time or was just too involved with the daily routine of living (going to work, mowing the lawn, cooking dinner, etc.) and by the end of the day I didn't feel like writing. Sometimes these routine periods can last for days at a time.

November 11, 1991 (pgs. 380 - 384) - Bonnie finally got to see the Florida Keys from the air today. I was signed up to fly N49464, but it had a dead battery, so I took N29EF (a C-152). The wind was calm when I took off from Runway 27R at North Perry Airport. I headed southeast to the beach. The air was hazy and slightly bumpy. Because of the haze, colors only showed up when looking straight down at the ground, then everything kind of blurred into the haze all around. In fact, when I got down to Key Biscayne via the shoreline, the bay and sky merged into the haze and I had to fly by attitude gyro, except to look out at the water below me on either side. Finally, Elliott Key emerged through the haze. While over the bay, I climbed from 1,000 feet up to 2,000 feet. I flew on the east side of Elliott Key, keeping the island on Bonnie's side. I pointed out the places we've sailed to around the bay, then Ocean Reef Club's private airstrip, Card Sound Bridge, Gilbert's Marina, and John Pennekamp State Park. I then crossed over Key Largo to fly down the west side of the island chain so that the islands were on my side. I assured Bonnie that the entire flight back would be from her vantage point. About this time, the air became increasingly bumpy. Next came Tavernier, Plantation Key, Islamorada, Upper & Lower Matecumbe Keys, and Long Key, at which point I called up Marathon on 122.8 to get the surface winds. The active was Runway 7 with a left-hand traffic pattern out over the water. I continued on while descending down to 1,000 feet. When I had Marathon Airport in sight I called Miami Flight Service to cancel my flight plan, then called Marathon to advise all traffic that I would be entering a left downwind for Runway 7. Another plane (C-150) was in the pattern doing touch-and-goes, so we continually communicated our positions and kept sight of each other. I entered a downwind when the other plane was on a base leg. Then, just as I was about to turn base, a corporate Westwind overflew me at about 1,500 feet and called that he was on a left downwind for Runway 7. When he started turning base about a half mile directly in front of me, I told him I would do a 360 degree turn to the right, re-enter the downwind and land after him. In the meantime, the other plane had taken off again and was climbing on the crosswind leg. I entered the downwind again, then turned base and final and set up a nice approach for my first island landing. The landing was solid. Then I had a long taxi to the ramp. Once at the FBO, I was guided to a tie-down spot and picked up by a golf cart for a ride to the terminal building. I spoke to the owners of the FBO and they stocked 6 copies of "Black Falcon", which made this flight my first successful business trip using a general aviation aircraft. Then the owners let Bonnie and I use their car to go to lunch. We drove to a waterfront grill. Bonnie had an oyster sandwich and I had an incredibly delicious hamburger with homemade french fries. Then it was back to the airport for the return trip. Oh, by the way, I got to meet the pilot of the plane that was doing touch-and-goes. He's in his 70s. He's retired and living in the Keys and he rents a plane once a month just for fun. I told Bonnie that if I ever live to be that old, I hope I'll still be flying, too. Because Marathon Airport is an uncontrolled airfield, pilots are suppose to report their positions and intentions over a common unicom frequency. During my run-up at the hold bars for Runway 7, I heard a pilot calling that he was passing overhead the southwestern tip of the island at 600 feet, so I knew to start looking out for traffic as soon as I got airborne. And I spotted him off to my right at around 600 feet as I was climbing out. Bonnie spotted him, too. He was flying a Piper Cherokee parallel to us. Then he turned left into a wide 180 degree turn, so I swung to the right to pass behind him. Then I continued climbing up to 1,400 feet. One thing was for sure - the haze had lifted, giving us a beautiful view of the Florida Keys for about 20 miles in all directions. The colors of the various bodies of water were absolutely breathtaking. Bonnie couldn't believe how clear the water was and how shallow it looked. We both looked for porpoises, but didn't spot any. Over Long Key a single-engine Cessna flew directly above me at around 2,000 feet in the opposite direction, then a short while later a C-402 flew parallel to me about a half mile to my left, but he was twice as fast, so I soon lost him. For this leg, all the sights were on Bonnie's side of the plane. She saw Duck Key, the island where we spent last Valentine's Day. And she saw Bud & Mary's Boat Yard, where we once rented a small Boston Whaler for the day. And, of course, she spotted Holiday Isle, her weekend hangout from her bachlorette days. And Plantation Key Yacht Harbor, where we saw an art show last year. And on and on with all the landmarks she had previously seen only from ground level. The perspective is so much different from the air. When we were over Biscayne Bay, I looked over at Bonnie and she was fast asleep. But I didn't know she was sleeping. I figured that maybe she was airsick and just keeping her eyes closed so she wouldn't have to look out. She is afraid of heights. So I asked her if she was okay. That's when I found out she had been sleeping, because she woke up. She said that the drone of the engine had lulled her to sleep. At that point, downtown Miami came into view in the distance. What a difference a few air miles can make. First, there's lush sub-tropical islands and boats beneath me and the next thing you know, I had skyscrapers and urban sprawl out in front of the windshield. I prefer the former. The rest of the flight was the usual shoreline ride north to Gulfstream Race Track, then a turn towards the west above the numerous 1,000+ foot radio and television antennas in the local antenna farm on the final leg towards North Perry Airport. I made a good crosswind landing on Runway 36R, then taxied back to the ramp. Total flight time was 2.7 hours for the 5 hour rental block. Bonnie said she really, truly enjoyed the flight.

February 15, 1992 (pgs. 462 - 463) - Valentine's Day (and night) went off without a hitch. The day started off with breakfast at a little cafe on Hollywood Beach. Then I took Bonnie to Coconut Grove, where we walked around all the specialty shops until we came to the store where I had ordered a special V-Day picnic basket especially made up for Bonnie. She saw the basket sitting on the counter, but didn't know it was for her. When the lady who worked at the store told Bonnie it was for her, she started to cry. Each item was individually wrapped in red and silver cellophane paper (wine, bread, cheese, salami, various fruits, glasses, corkscrew, but they forgot the grapes and strawberries I had ordered). Bonnie and I took the basket and walked to Peacock Park, found a shady spot under a tree that overlooked Biscayne Bay, laid out a blanket and had a very romantic lunch, with jazz on the radio. Afterwards we layed down on the blanket and cuddled. Then we came home and the dozen long-stemmed roses I had ordered were waiting for Bonnie at a neighbors house. Then we looked in "The Miami Herald" to read the "True Romance" article they had promised for Valentine's Day. They had asked readers to send in examples of true romantic experiences and Bonnie was one of 261 people who had responded. She wrote about the time I gift-wrapped myself for her and also the time I mowed a heart and the words "I Love You" into the front lawn. Well, they picked the lawn mower episode and published Bonnie's writing in yesterday's article. She was one of about a dozen people whose story got published. Bonnie, of course, was tickled over this. Then Bonnie and I took a shower together and then I made her stay in the bedroom while I set up the dining room for a surprise candle-light dinner. I then had to drive to our favorite restaurant, where I had arranged to pick up two take-out dinners. When the meals were prepared on the table, I let Bonnie come out of the bedroom. We ate a nice dinner together, then made love.

September 26, 1994 (pg. 1000) - Hey, I've hit page 1000 of this journal.

March 8, 1996 (pgs. 1366 - 1367) - It's unbelievable! Great news! Walter J. Boyne, world-renowned aviation author, wrote me a letter saying that he will write the foreword to "How To Write, Publish & Sell Your Own Aviation Books". I am so excited!

July 14, 1997 (pg. 1686) - I finished the first draft of "How To Write, Publish & Sell Your Own Aviation Books" last night (actually early this morning). Because I wrote this book on my computer, it should only take one rewrite.

APPENDIX B

BIBLIOGRAPHY & FOOTNOTES

Bibliography

A Hostage To Fortune, by Ernest K. Gann, 1980, Ballantine Books, original edition 1978, Alfred A. Knopf.

Author, Pilot Martin Caidin Dies, by Editorial Staff, May 1997, AOPA Pilot.

Bizplane People, by Editorial Staff, July/August 1992, Aviation International News.

Book 'Em, by Tom Matthews, July 1993, TWA Ambassador, original edition n.d., Newsweek.

Checklist: A Review Of The Latest Airliners Literature, by Editorial Staff, Summer 1989, Airliners.

Do-It-Yourselfers Carve Out A Piece Of The Publishing Pie, by Doreen Carvajal, April 28, 1996, The New York Times.

EAA/TSI Amateur-Built Aircraft Investigation Course, by Earl Lawrence, February 1997, Sport Aviation.

Evans' Novella Tugs Families' Heartstrings, by Deirdre Donahue, December 14, 1995, USA Today.

Florida's Publishers, by Ted Reed, November 13, 1989, The Miami Herald/Business Monday.

How To Bulletproof Your Manuscript, by Bruce Henderson, 1986, Writer's Digest Books.

Jurors' Vanity Press Review: Publisher Defrauded Authors, by Ronald Sullivan, n.d., The New York Times.

Keys Figure Prominently In New Adventure Novel, by David Hawkins, December 28, 1988, Florida Keys Keynoter.

New Novel Covers South Florida Drug Scene, by Milt Baker, February 1989, Waterfront News.

Put Pen To Paper These Days, You'll Be A Published Author, by Cynthia Crossen, January 10, 1997, The Wall Street Journal.

Small Has Never Looked So Beautiful, by Eva Pomice, January 9, 1989, U.S. News & World Report.

Small Publisher's Destiny, by John P. Dessauer, October 1987, Small Press.

Summer Editions: Books And Videos To Tide You Over When Storm Warnings Fly And Your Boat Can't, by Jane Tracy, June 1989, Cruising World.

The Autobiography Of Mark Twain, arranged and edited by Charles Neider, 1959, Harper & Row Publishers, original edition by Mark Twain, 1917, The Mark Twain Company.

The Basic Art Of Submitting A Book Manuscript, by Michael Seidman, October 1990, Writer's Digest.

Bibliography - Continued

The Business Of Being A Writer, by Stephen Goldin and Kathleen Sky, 1985, Carroll & Graf Publishers.
The Self-Publishing Manual, by Dan Poynter, 4th Edition 1986, Para Publishing.
The Self-Publishing Handbook, by David M. Brownstone and Irene M. Franck, 1985, Plume Books.
Twice-Told Tale, by Debbie Sontag, August 15, 1991, The Miami Herald.
Writer's Market 1992, by Editorial Staff, 1991, Writer's Digest Books.
Writing Into The Year 2000: Your Future In Book Publishing, by Russell Galen, January 1990, Writer's Digest.

Footnotes

1. The Self-Publishing Manual, p. 24.
2. Writing Into The Year 2000: Your Future In Book Publishing, p. 25.
3. Small Publisher's Destiny, p. 82.
4. Writing Into The Year 2000: Your Future In Book Publishing, p. 25.
5. Ibid., p. 25.
6. Ibid., p. 29.
7. Bizplane People, p. 114.
8. Twice-Told Tale, n.p.
9. Writer's Market 1992, n.p.
10. Ibid., n.p.
11. The Basic Art Of Submitting A Book Manuscript, pp. 30-32.
12. Writer's Market 1992, n.p.
13. Small Has Never Looked So Beautiful, p. 49.
14. Small Publisher's Destiny, p. 82.
15. Small Has Never Looked So Beautiful, p. 49.
16. Florida's Publishers, p. 1.
17. Ibid., p. 15.
18. Ibid., p. 15.
19. Jurors' Vanity Press Review: Publisher Defrauded Authors, p. 1.
20. The Business Of Being A Writer, p. 4.
21. Ibid., p. 4.
22. The Self-Publishing Manual, p. 5.
23. The Autobiography Of Mark Twain, pp. 153-154.
24. Ibid., pp. 225-228.
25. Ibid., pp. 228-236.
26. Do-It-Yourselfers Carve Out A Piece Of The Publishing Pie, p. 22.
27. Evans' Novella Tugs Families' Heartstrings, p. 2.
28. EAA/TSI Amateur-Built Aircraft Investigation Course, p. 66.
29. A Hostage To Fortune, p. 463.
30. Summer Editions: Books And Videos To Tide Your Over When Storm Warnings Fly And Your Boat Can't, p. 91.
31. A Hostage To Fortune, p. 463.

Footnotes - Continued

32.New Novel Covers South Florida Drug Scene, p. 21.
33.Checklist: A Review Of The Latest Airliners Literature, p. 55.
34.A Hostage To Fortune, p. 463.
35.Put Pen To Paper These Days, You'll Be A Published Author, p. 61.
36.How To Bulletproof Your Manuscript, p. 33.
37.Ibid., p. 34.
38.The Self-Publishing Manual, pp. 53-59.
39.Ibid., p. 96.
40.Ibid., p. 96.
41.Ibid., p. 99.
42.Ibid., p. 99.
43.Ibid., p. 118.
44.Ibid., p. 111.
45.Ibid., p. 111.
46.Ibid., p. 103.
47.Ibid., p. 103.
48.Ibid., p. 103.
49.Ibid., p. 104.
50.Ibid., pp. 105-107.
51.Ibid., p. 106.
52.Ibid., p. 119.
53.The Self-Publishing Handbook, n.p.
54.Book 'Em, p. 24.
55.Keys Figure Prominently In New Adventure Novel, p. 28.
56.New Novel Covers South Florida Drug Scene, p. 21.
57.Checklist: A Review Of The Latest Airliners Literature, p. 55.
58.Author, Pilot Martin Caidin Dies, p. 31.

Turn the page for *Black Falcon* excerpts!

APPENDIX C

BLACK FALCON - SELECTED EXCERPTS

You have just read the true story about writing and publishing *Black Falcon*. Now is your chance to read actual excerpts from this thrilling action-adventure novel. Please note that strong language is used throughout the text of *Black Falcon*. As a courtesy, I have deleted the f-word from the following excerpts.

Hi,
Please buy "Black Falcon"
so Daddy can get his airplane.
Love,
me Jennifer

BLACK FALCON

Early morning in mid-February

The gravel-faced captain maintained a firm grip on the Lodestar's control wheel as he hand flew the twin-engined aircraft just fifty feet above the Florida Straits. He was thinking about the rip-off, wondering if it would go down as planned, wondering if he'd have enough guts to pull the trigger of his .357 Magnum. He had never killed a man before. Would it be the same as killing a deer? He wondered a bit more, then put his mind back on flying the airplane.

After scanning the panel, the captain turned his gaze out the windshield. A full moon hung low above the horizon as it made its nightly descent against the western sky. Below, the waves' crests sparkled in the moon's effervescence, creating an eerie, yet tranquilizing feeling for the pilot.

The Lodestar's copilot, meanwhile, was paying no particular attention to nature's artistry, although he, too, was looking out the aircraft's windshield. His eyes were busy darting around, searching in the distant early morning darkness for a certain man-made landmark. He was sure his navigational calculations were correct. Or were they? Maybe he had figured the winds wrong.

Maybe they were stronger than anticipated. Straining his eyes, he continued peering out through the windshield.

Finally, after what seemed like half an hour, the Middle Keys came into view. "Over there!" the copilot shouted, pointing about fifteen degrees to starboard. In the distance, the tiny islands could be seen as dark voids against the water's shimmering moon-glow. Pointing again, this time a few degrees to port, he added, "The Seven Mile Bridge is over there. That's our mark. Marathon's the long island off to the right of it. Damn...right on the money!"

The captain acknowledged the sighting with an almost undetectable nod of his head. He checked his watch. It was 4:37 in the morning. Reaching down into his flight bag, he felt for his gun, reassuring himself that it was still there, ready to be used at a moment's notice. Then, not wanting his uneasiness to show through, he slowly withdrew his hand from the flight case, leaving the .357 braced between his Jep charts and a thermos. Nervously, he scratched at his chest.

"Three-one-five degrees oughta get us over the bridge," instructed the copilot as he mentally plotted a new course for the Lodestar.

"Three-one-five it is." Gently, the captain initiated a turn towards the bridge.

The copilot was checking a Miami sectional chart with a red-lensed flashlight. "There's powerlines along there. Along the bridge, I mean. I'm not sure how high, though." The high-tension lines were not indicated on the chart. The experienced aviator was speaking strictly from memory.

The captain heeded the warning by influencing the heavily loaded airfreighter up another hundred feet. Might as well play it safe. No point taking any chances at this stage of the game, he mused in the back of his brain.

Things were deadly enough as they were with the tons of marijuana in the cabin.

Still scrutinizing the chart, the copilot's eyes were drawn to an area of restricted airspace located about eighteen nautical miles to the west of the Seven Mile Bridge, directly over Cudjoe Key. The airspace was home to *FAT ALBERT*, an unmarked tethered balloon carrying "Look-Down" Doppler radar. Although the balloon was used for spotting low-flying aircraft trying to sneak into the country below the coverage of normal ground-based radar, the right-seat pilot was not worried. Falcon had taken care of everything. Falcon always took care of everything. It was imperative. The organization's survival depended on it. Falcon, after all, controlled one of the largest marijuana importation rings in the United States.

It had been easy for Falcon to get his pilots the window through the government's radar net. All it took was money. Lots of money. Cash money. Everybody had their price. For some, it was $5,000. Others held out for larger amounts—$15,000, $25,000, even $50,000 depending on the value of the information.

That had been the price of tonight's intelligence. Fifty thousand green ones. In exchange for the cash, Falcon had received a message. Short and simple, it had read *FAT ALBERT INOPERATIVE TONIGHT*. But those four words meant a great deal to Falcon and to the two pilots flying the Lodestar over the Florida Keys. It meant safety. It meant free skies. It meant profits.

The message told them that the Aerostat radar balloon at Cudjoe Key was down for maintenance. Images of low-flying aircraft were not being sent to radar screens in Miami—not tonight. The Lockheed Lodestar was free to go about its clandestine business as planned, flying straight towards the bridge, just south of

Marathon.

The Overseas Highway became clearly visible as the aircraft drew closer to it. Then, as the Lodestar passed over the bridge, the captain put the plane through a series of Dutch rolls. Together, both flyers kissed the Atlantic good-bye and welcomed in a new body of water—the Gulf of Mexico.

"If this ain't like the government right up its ass, then I don't know what is!" exclaimed the copilot, which he followed with a hearty laugh. He then turned around in his seat to face a third man sharing the cramped cockpit with himself and the captain.

The man was of a darker complexion than the pilots—not negro, but Latin. A solid gold bracelet was clasped around his right wrist, inscribed with the name *GUSTAVO* in large bold letters. A diamond-studded gold Rolex adorned his left wrist. He was Falcon's trusted courier—the man responsible for seeing that the load arrived safely in Florida.

If Gustavo had spoken five words of his broken English during the entire all-night flight up from Colombia, it would have been a lot. But the twenty-four year old Colombian national remained quiet in his seat as the Lodestar droned on towards its destination.

Buckled snugly to the worn leather cushion of the cockpit jumpseat, directly behind the flight crew, Gustavo was feeling edgy. He sensed something, although he didn't know what. Not wanting anything to go wrong, he rested his nimble fingers on a MAC-10 automatic pistol, which was tucked awkwardly in the waistband of his pants. The powerful weapon was much too big to be carried the way he had it, but that's where he kept it. Pure Latin machismo.

The copilot said to Gustavo, "That intelligence crap sure pays off nicely, don't it?"

For no apparent reason, except maybe for his extreme edginess, Gustavo came alive from the flyer's remark. He drew the MAC-10 from his pants and pressed the weapon against the copilot's skull. "My friend, you flys only. You no talks to me." Then, just as fast as the automatic pistol was drawn, it was quickly and efficiently pushed back into Gustavo's waistband. A grin was plastered on his face. "I scares you?" he asked.

There was no answer from the copilot. He was too terrified to even open his mouth, let alone speak.

"I scares you." This time it was a statement of fact. "You turns around and flys dis plane."

The copilot did just that, turning away from Gustavo and setting his eyes once again upon the aeronautical chart laid out on his lap. From there, he gazed up at the instrument panel where he scanned a breadth of dials and gauges, half of them no longer working from old age. Then his eyes quickly glanced back down at the chart, then out the windshield and finally, he settled his sight on the lighted panel. He was now as nervous as the captain. He thought about the rip-off. Damn, he wished the captain would give him the signal already. He wanted to get it over and done with.

Fifteen minutes slipped by before the captain reached forward and turned a worn black knob on the panel, dimming the instrument lights. Although this made it somewhat difficult to read the gauges, it was a necessary part of the plan.

It was also the signal the copilot had been waiting for. It was time. The copilot looked over towards the captain. "Final heading is three-four-five. That oughta bring us straight over Charlotte Harbor. We'll follow the ADF in from there."

The Lodestar's captain banked the aircraft into a shallow turn to the north, then leveled off the plane on the correct heading of 345^{0}. The scheduled destination, as arranged by Gustavo, was a rarely used dirt road near the coastal town of Punta Gorda. The brazen pilots, however, were not counting on landing there. They had another site in mind.

Over the next few minutes, the captain allowed the Lodestar to deliberately drift off course...past 347^{0} on the magnetic compass...past 350^{0}...past 355^{0}. He finally fixed the course on 358^{0}. Straight ahead lay rural LaBelle, a quiet farming community southwest of Lake Okeechobee. It was there that the pilots were planning to set the plane down on a freshly plowed cucumber field, some twenty-five miles to the east of Punta Gorda.

For their plan to work, it was a must that Gustavo not get a good look at the dimly-lit instrument panel. It was imperative that Gustavo not suspect a thing. Both pilots knew they had to act as natural as possible for the rip-off to go down successfully. And if it did, Falcon's profits would be theirs to split. Instead of two million dollars for Falcon and $100,000 for each of them, it would be two million for them. It was a risky venture, but worth every bit of the gamble.

For the next half-hour, the cockpit remained relatively quiet—almost peaceful. The .357 Magnum was continually on the captain's mind. He kept anticipating its usage. He still wasn't sure if he could kill a man in cold blood. Finally, he knew the time had come. With his left hand remaining firmly on the control wheel, he reached into his flight bag for...

...the flash was blinding, the echo deafening as Gustavo fired off several rounds from his MAC-10 into the captain's heart from behind. Blood and body tissue splattered against every inch of the stale cockpit.

The copilot, although sickened by the horrendous sight surrounding him, instinctively grabbed for the wheel and took control of the aircraft. In less than three seconds, the Lockheed Lodestar had descended almost into the sea. Gradually, the flyer regained most of the plane's lost altitude. But things were not looking good for him. He could feel the still-hot gun barrel of the MAC-10 being pressed smartly against the back of his neck.

With authority, Gustavo told the copilot, "Now you lands where I says. Now you flys to Punta Gorda."

The frightened aviator had no choice but to follow Gustavo's demand. He banked and turned the plane to the left, setting the airfreighter on a course for Punta Gorda.

A short while later, over Charlotte Harbor, a radio signal was picked up on the Automatic Direction Finder from the landing site's portable ADF station. The copilot tracked inbound on the invisible electronic beam until he spotted car headlights marking the dirt road in the distance. With two miles to go, he eased back on the power levers, lowered the flaps and then the gear. All lights were indicating green. Touchdown was moments away.

Sweating profusely, the copilot fought as best he could to maneuver the slow-flying airfreighter through the dark nighttime sky, but the controls felt sluggish in his hands, preventing any bit of real precision. Considering the circumstances, though, he was doing a hell of a job handling the plane.

Seconds later, the Lodestar hit the ground with a loud thud of rubber on dirt. The engines were shut down as the plane came to a full stop in the middle of the road. Several four-wheel drive vehicles were standing by with ground crew, all of them eager to begin off-loading the bales of pot from the late arrival.

Meanwhile, up in the cockpit, Gustavo shot and killed the copilot by piercing the flyer's brain with a round of bullets from his MAC-10. Then, to satisfy his ego, Gustavo fired off two more shots—this time with his Nikon 35mm. Once developed, the pictures would be added to the Colombian's grotesque photo collection of lives he'd eliminated in the line of duty.

Before exiting the cockpit, Gustavo had one final chore to complete. Reaching forward, he yanked the black falcon charms from around the necks of both dead flyers, removing the only evidence linking the crime to the organization.

With his job done for the day, Gustavo left the aircraft and walked towards a vehicle waiting to rush him back to Miami, all the while contemplating his next assignment—finding replacements for Falcon's two dead crewmen. He knew the task would not be a difficult one, for there were so many fools in this world. So many fools.

November - Friday Morning

Robert Jensen stared out at the aging Douglas DC-4 from the rear seat of the black and white taxi as the cab pulled onto the ramp at Houston Intercontinental Airport. The tramp airfreighter looked uninviting. Almost as uninviting as life itself.

For someone who had only been around for twenty-nine years, Rob Jensen had seen a lot—much of it not good. His mind and body were tired, his emotions distraught. Many times he had wanted to pack everything in and say, "To hell with it all." End his life. Suicide.

But he never did. Something always came up that needed doing. A left-seat flying job. A favor for Maureen. Something. And so, just barely, Rob kept on with his life—breathing, flying, maintaining his six-foot-two-inch body the best he could.

It was possible, of course, that deep down Rob really didn't want to die. Perhaps his thoughts were just bent towards the extreme. Not that it mattered much. The fact was, the suicidal tendency was there. Rob knew it. The deadly act was always roaming the far reaches of his mind, occasionally creeping forward.

Through the cab's dirt-tinted windows, Rob glanced

up at the deep blue sky. Above, there was nothing but a few wispy, feather-like cirrus clouds suspended high in the southeastern quadrant.

Rob rolled down the side window and breathed heavily. The air was crisp and light with no hint of smog. It was a perfect day for flying. Perfect in all respects but one—Rob didn't feel like flying.

Despite his personal feelings, however, he knew he had to fly. It was his job to ferry the beefy airfreighter down to Opa Locka Airport in South Florida. The black falcon charm around his neck reminded him of it. The man named Falcon demanded it. And Gustavo would most definitely be at the airport to make sure of it.

Rob cursed the day he involved himself with the organization. He cursed Falcon. He cursed Gustavo. He cursed the whole damned world—except for Maureen. Why had Gustavo's "come-on" of making a hundred grand for one night's work been so tempting?

It was the bass, Rob thought. That was the answer. Bass and the goddamned cargo operators he'd been flying for, bringing home less than $15,000 a year after taxes. That's why he'd done it. That's why he'd fallen for Gustavo's lines. He was going to make the one dope run, take his cash and leave the pressures of society behind. He was going to sell his mobile home—buy a cabin out in the sticks somewhere. He was going to spend more time bass fishing. that noise, though. Falcon had forced Gustavo to shove more runs down his throat, with the constant threat of death being the retainer.

And so, here he was in Houston, Texas, almost nine months to the day. He still didn't own the cabin in the woods. The tin-can trailer was still in his name, mortgaged to the hilt. At least he'd kept his promise of doing more fishing. And he had a spanking new, top-of-the-line bass boat sitting in his driveway, paid for with cash—dirty money. Rob cursed Falcon again as the taxicab pulled closer to the plane that needed ferrying.

Riding in the back of the cab with Rob, sitting motionless, was Nick Donovan, Rob's copilot for the flight to Florida. Twice Rob's age and three times as scruffy, Nick also didn't give a damn about piloting the DC-4 parked out there on Houston's ramp. His reason, though, wasn't the same as Rob's, for Nick put up with flying for Falcon. The work was exciting and the high pay supported his many habits. Nick's reason was simply that his mind was someplace else, lost in a fantasy world as he watched a Boeing 747 gracefully liftoff from the runway, probably headed for some faraway destination. It should have been him strapped in the spacious cockpit of that jetliner, his jacket sleeves proudly displaying four gold captain's bars. It should have been him—but it wasn't. Nick had blown his chance long ago when he had let alcohol and a non-conformist attitude put a damper on his career with a major South American airline.

Now, here he was pushing sixty years of age with a graying beard and wrinkles, still copiloting the same junk he flew back in the 1940's and 50's. The same types of airplanes that had cost him his left foot in the crash down in Brazil—old, radial-engined transports. The DC-4's ...DC-6's...Lodestars...Beech-18's. They were all basically the same. Flying death traps. That's what they were. Flying death traps. But they were all Nick had, so he stuck with them.

When the taxi finally pulled up next to the airplane, Rob handed the driver a twenty dollar bill for the $4.50 fare. "Keep the change," he said generously as he and Nick exited the cab.

The driver stuffed the bill into his shirt pocket, lit a cigarette and sped away, leaving the two pilots alone with their airplane.

But Rob and Nick were not alone. From across the ramp, a short man with jet-black hair approached them. As he neared, it was apparent he had a gun stuffed into the waistband of his pants. The man was no stranger to the two pilots.

"Buenos dias, my friends!" It was Gustavo in all his splendor. A racing jacket matching the color of his dark, wavy hair was unzipped almost to Gustavo's waist. Diamond-studded cowboy boots and a pair of denim jeans finished off the outfit. Gustavo's usual amount of heavy gold jewelry dangled from his body, including the chain around his neck with the black falcon charm on it. "How flight was?" he asked in his broken English, referring to the commercial flight the two Floridians had taken to get to Houston.

" male stewardesses," Rob commented defiantly, his crude undertones intended. He never did adjust to the airlines' current trend of presenting male flight attendants to the flying public. After all, what had been so terrible about a nice pair of thirty-eight double-D's walking up the aisle—or at least being served an ice-cold Budweiser by a pretty face with a set of perky A-cups? Rob just couldn't understand the airlines' thinking.

Nick also answered Gustavo's inquiry by grumbling something unintelligible under his breath. Probably something about the change-of-plane layover in Atlanta. Always in Atlanta. Nick hated flying as a passenger on commercial flights. It made him uncomfortable not having control of the aircraft in some way.

Today's flight had been especially torturous for Nick. He'd gotten stuck sitting next to an old lady who insisted on telling him all about her wonderful grandchildren. All eleven of them. And to top if off, the flower-hatted lady had had a wallet that contained a non-ending gallery of family photographs as proof to her borish chatter. Nick hated flying as a passenger.

When the greetings were over between the three men, they got down to business. Nick went off to do the preflight inspection, while Gustavo and Rob discussed the aircraft's overall condition.

"The maintenance logs?" Rob asked quizzically.

Gustavo pointed towards the cockpit, indicating that the books were already inside the airplane. "Is no problems," he insisted. "Plane is good."

"Yeah, that's your opinion. I still wanna see the logs." Rob was eyeing a puddle of grimy oil on the ground beneath the number two engine. It was normal for radial engines to drip oil, but the number two definitely had a major leak. The other three engines looked fine to Rob, with each of the powerplants having only minor puddles to account for.

Rob walked over to the suspicious number two engine, mounted inboard on the left wing. Gustavo followed him over. Together, they examined the evidence.

"I don't know, Gustavo. You may be wrong. I think this plane's got problems. This leak could be serious...or it could lead to something serious." Rob reached up and ran his finger along the bottom of the engine cowling. He thought for a moment as he wiped his oily finger on his shirt. "It could be bad rings...or guides. Maybe we've got a cylinder head that's shot all to hell...who knows? I guess it could be nothing, too. I'll have George open her up tomorrow and take a look." Turning away from the engine and towards Gustavo, Rob asked, "When's our next run to Colombia?"

Gustavo pulled out a folded piece of paper from his left hip pocket. He unfolded it and handed it to Rob. "Here, my friend." It was the logistics sheet for Falcon's next drug flight. "We flys Saturday," Gustavo pointed out.

"No way! That's tomorrow!" Rob exclaimed, shaking his head from side to side.

Gustavo laughed. "I gives you times. We no flys 'til next Saturday."

"Oh...okay, then." Relief was in Rob's voice. "I guess George can do something about the leak by next Saturday. That gives him a week."

"*Si.* George is good mechanics."

"That depends who you ask," Rob responded with sarcasm.

"Ah, my friend. I remembers. You no likes each others."

No comment from Rob. To him, Falcon's mechanic was nothing but a lazy waste of human flesh. Rob, of course, had no say in the decision. For some reason, George was Falcon's choice as the organization's line mechanic. Rob turned his attention back to the logistics sheet, studying the paper attentively.

Gustavo sensed what Rob was looking at and put forth, "You knows where is this Hato Nuevo is? Where we flys to Saturday?" Gustavo's English was getting worse with time. He never was a very good scholastic achiever.

Rob looked up. "I've got an idea of where it's at. I know it's somewhere inland from the coast of the peninsula...somewhere east of Barranquilla...almost near the Colombian-Venezuelan border, I think."

"*Si. Peninsula de la Guajira.* You knows where is this is then. You no loses us."

"I no loses us, Gustavo," Rob kidded. "What about the landing site? Any details on it yet? That last one was pretty marginal."

Gustavo looked over his shoulder to see how Nick was coming along with the preflight inspection, completely ignoring Rob's inquiry about the destination's landing field.

"What about the airstrip?" Rob asked again.

Gustavo turned back to Rob. "I picks good strip. Plenty feets for airplane."

"Yeah, sure." Rob's sarcasm glared through once again. He knew that Gustavo didn't know shit when it came to picking airplanes and that the Latin knew about the same when it came to landing sites. And yet, like George was the mechanic, Gustavo was Falcon's procurement man. Ironic, Rob thought, that as always, the next drug flight was doomed to count heavily on luck. Rob figured he'd probably be able to handle it if the organization was run three-quartered-assed, but half-assed just didn't buy it for him. "Anything else you wanna let me in on?" Rob wanted to know.

"No. I sees you backs at Opa Locka, my friend." Gustavo turned and walked away from the airplane.

When the Latin had gotten about thirty yards distance, Rob called out, "Hey! Is this thing fueled?"

"Si!" Gustavo shouted back as he continued walking.

But Gustavo was called at again—this time by Nick. "Man, you suck, Gustavo! This plane is crap!"

Gustavo's reaction was like a bomb going off. He couldn't have gotten any hotter had there been a nuclear detonation on the field. He came storming back to the airplane with hatred in his eyes. Although he was much smaller physically than Nick, he shoved the burly copilot about three feet backwards using his outstretched fore-arms. He shouted "Look, old mans! I takes you out one days. You keeps talks like this, I kills you."

Nick spoke loudly, too, egging Gustavo on. "Man, without that gun of yours, you'd be nothing but chicken shit."

Gustavo's dark eyes glared forcefully. He barked, "*Maricon*!" He knew Nick understood Spanish. Then, without any regard for the possibility of on-lookers, Gustavo withdrew his gun and threw it wildly to the ground. With nostrils flaring and both hands clenched into muscular fists, he shouted at Nick, "Come. Charge me. I wait for months this fight."

Being an avid barroom brawler, Nick realized his opportunity at getting in the first punch. He lunged forward, landing a few violent blows into the Colombian.

Gustavo staggered away momentarily. He put his hand to his mouth, where he had taken a particularly hard punch from Nick. His lower lip was bleeding. Gustavo soon regained his composure, though. Coming back with vengence, he surprised the flyer with his own style of street fighting, learned a decade ago when he was a member of a teenage youth gang in Cartegena.

Rob kicked the gun—a MAC-10—out of reach of either man, but did nothing to stop the altercation. He stood back and watched the fighting with detached bemusement. He knew the fight was overdue. Each man had been chewing at the other's ego for too long. They were venting anger. Buy why did they have to do it out in the open? What if somebody drove by and spotted the gun? It would raise suspicions—something none of them needed.

The fight lasted another seven minutes before Nick and Gustavo simultaneously called it a draw. Both were out of breath. Both their faces showed battle wounds—open cuts, puffiness, oozing red blood. They stood eye to eye. Their feud was not over. Not just yet.

Gustavo walked away, leaving the two crewmembers to ready Falcon's airfreighter for the flight from Texas to Florida.

The sound of thundering piston engines could be heard reverberating through the morning air as Rob taxied the DC-4 out to Houston's Runway 32. He held the aircraft short of the runway as a Lockheed L-1011 began its takeoff roll down the 12,000 foot slab of concrete.

After a short delay, an air traffic controller said over the radio, "November-Two-Five-Three-Kilo-Yankee cleared for takeoff, runway three-two. Wind three-four-zero at seven. Caution wake turbulence, departing heavy. Immediate right turn to zero-niner-zero VFR approved." Terse and to the point. Each word carried important information to the pilots of N253KY.

"Three-Kilo-Yankee is rolling," Nick said into the mike, acknowledging the radio transmission.

Takeoff power was applied. The aircraft moved forward, constantly building speed. The load was light—no cargo, no passengers.

Minutes later, Rob and Nick were leveled off at 7,500 feet, passing over the Gulf of Mexico after crossing Galveston Bay. A mile long oil slick could be seen on the water's surface in the distance off the airfreighter's left wingtip—modern man's contribution to the environment.

Rob engaged the autopilot, then looked over at Nick. "What made you let loose like that back there in Houston?"

"That machismo little called me an asshole and nobody calls me an asshole and gets away with it." As he spoke, Nick picked dried blood from his full growth of facial hair.

"I never heard him call you an asshole."

"Man, that's 'cause you don't speak Spanish. *Maricon*...asshole. Same shit either way."

"Yeah, well, if you keep razing Gustavo like that, you're gonna end up as just another snapshot in his album." Rob waited for a reply. He knew he'd get a good one.

"I hate that Colombian faggot!" Nick fumed, his wrinkles becoming more apparent in his fit of anger.

"Go ahead and hate him all you want," Rob said, in agreement with his copilot's feelings. "But shit, Nick, do it behind his back. Don't flaunt your hatred in the man's face."

"Hey, like I tell the ladies...if you got it, flaunt it. And, man, I got it."

And that was all the two of them said to each other for the next several hours. Then, somewhere over Central Florida, the musty cockpit again became alive with conversation.

Rob was the one who broke the silence. "Shit, I'm starved. A burger, fries and chocolate shake would go good with me right about now." When Nick didn't respond, Rob checked the panel clock. "It's three o'clock already. We won't be on the ground for another two hours. How about if we make an unscheduled pit-stop in Orlando? I'll buy."

"Man, you should've ate that airline food on the way in to Houston. It wasn't that bad. First-class stuff and all."

"You call rubber steak and powdered eggs *not that bad?"*

"Hey, I swallowed mine without puking."

"I swear, Nick. You must have a lead-lined stomach."

"Man, I had to eat. I was so hungry, I was starting to see God."

Pushing Nick's humor aside, Rob asked again, "Well, what about it? Do we go for Orlando?" He was rubbing his stomach to emphasize his suggestion.

"Nah. Let's just get this bird home. You'll live."

"We can tell Gustavo we had a mechanical," Rob offered, still trying to convince Nick of his idea.

“Hey, you’re the captain, man. Do what you want. I’ve already given you my worthless opinion.”

Rob knew Nick was right about not landing in Orlando, so he shut his mind off to food. Silence once again filled the cockpit. Although Rob’s mood was relaxed, a thick air of invisible tension was building up inside of him as the aircraft continued southward, flying above acres of neatly rowed citrus groves.

A few minutes later, Rob spurted weakly, “I’m getting out.”

“What the hell you talking about?”

“From Falcon. From the organization. I’m getting out...after our next run.”

“Man, you’re   crazy. You know you won’t do it.”

“   you, Nick.”

“Yeah,   me!” Nick laughed a gutteral laugh that came from deep down inside his diaphragm. He pulled at his beard.

“I’m serious. I’m leaving. I’m escaping. You should, too.”

“What for?”

“Cause we’ve been in this organization for too long. Falcon’s using us. You know it and I know it. We both went into this thing nine months ago to make one run, grab a hundred grand each and get out. But look what’s happened. Falcon’s already forced four more runs on us. And now this next one. You know when it’s gonna end, Nick? It’s gonna end when Falcon gets tired of us. He’s not gonna want us hanging around here forever, you know. And when the time finally does come that he decides to have Gustavo hire fresh meat, he’s gonna have us wasted...just like those bastards before us. That’s why we’ve got to make a break for it now...before Falcon breaks us.”

WRITER'S BOOK MART CATALOG

A Division OF
TRI-PACER PRESS

Reference Books For The Self-Publishing Industry

(Books by Chevy Alden are available at quantity discounts for resale!)

Black Falcon: (by Chevy Alden) This thrilling air-sea novel is a violent flashback to a time when South Florida's nighttime sky was a shadowy place to be...the early 1980s...the boom years for aerial marijuana smuggling. Drugs were cool...sex was loose...life was easy. Instant wealth was offered to any pilot willing to risk his life flying tramp airfreighters down to Colombia and back. One of those outlaw airmen was Rob Jensen, a physically strong, but emotionally unstable doper pilot who entered into a desperate struggle for survival against his distraught emotions, nature's elements and the deadly world of drug running. Follow along as this fast-paced, action-adventure novel takes you from a sleazy Miami bar to a salty Key West tavern, from the cockpit of a drug plane to a clandestine airfield in the mountains of Colombia, from a tropical Bahamian cruise aboard a private sailboat to a climactic action-packed ending on Fort Lauderdale's Venice-like canal system. Like so many pilots before him, Rob Jensen had learned the hard way that smuggling was a fool's game...a deadly game not to be reckoned with. This is the book that started it all for Chevy Alden! Get your copy today! $20.95 - Hardcover.

How To Get Published - Guaranteed: (by Chevy Alden) This award-winning self-help manual reveals the secret of guaranteed publication and teaches you the tricks-of-the-trade that Chevy Alden learned the hard way from trial and error. Topics include writing, editing, publishing, promoting and marketing. Written in an easy-to-understand format, this manual is enriched with more than 100 photos, figures, illustrations, documents, publishing forms and sample letters, plus a catalog filled with numerous publishing-related reference books. A must-read for every author who wants to learn the art of publishing from an insider's point-of-view. $21.95 - Trade Paperback.

How To Write, Publish & Sell Your Own Aviation Books: (by Chevy Alden) This self-help manual is an aviation version of Alden's award-winning reference book "How To Get Published - Guaranteed". A must-read for authors of aviation books. $21.95 - Trade Paperback.

How To Write, Publish & Sell Your Own Nautical Books: (by Chevy Alden) This self-help manual is a nautical version of Alden's award-winning reference book "How To Get Published - Guaranteed". A must-read for authors of nautical books. $21.95 - Trade Paperback. (This book will not be available until sometime in the near future.)

How To Promote & Market Your Own Books: (by Chevy Alden) This reference book is the sequel to Alden's three self-help manuals: "How To Get Published - Guaranteed", "How To Write, Publish & Sell Your Own Aviation Books" and "How To Write, Publish & Sell Your Own Nautical Books". Topics include free publicity, book reviews, advertising, personal appearances, speaking engagements, book displays, book shows, trade shows, specialty events, selling subsidiary rights, selling to bookstores, book wholesalers, book distributors and specialty stores, and a whole lot more. A must-read for authors who really want to sell more of their books. $21.95 - Trade Paperback. (This book will not be available until after the year 2000.)

1001 Ways To Market Your Books: (by John Kremer) This reference manual features over 1,000 ideas, tips and suggestions for marketing your books - all illustrated with real-life cases showing how other publishers have marketed their books - from planning marketing strategy to editing for promotional clout, from getting publicity to selling your books to specialty markets. This is the manual that separates the wannabes from people who are serious about making money with books. $27.95 - Trade Paperback.

The Self-Publishing Manual: (by Dan Poynter) This reference manual is one of the most authoritative guides to self-publishing - the classic reference manual for writing, publishing, promoting, marketing and distributing your books. It takes you step-by-step from initial idea through manuscript development, printing, binding, promotion and sales. $19.95 - Trade Paperback.

Is There A Book Inside You?: (by Dan Poynter and Mindy Bingham) This reference manual will teach you how to pick your topic, how to break it down into easy-to-attack projects, how and where to do research, how to improve your material and how to find writing partners. You will also learn a process that makes writing almost easy. This manual will show you how to write your book and how to get the help you need. Includes a list of writing resources. A Writer's Digest Book Club main selection. $14.95 - Trade Paperback.

Ordering Information

To place an order, send a complete listing of the books you want with the total dollar amount of your order. Please print your name and address clearly. Mail your book order to:

TRI-PACER PRESS
Writer's Book Mart Catalog
P.O. Box 840111
Pembroke Pines, FL 33084-2111

Method of Payment: Checks or Money Orders only *(payable to TRI-PACER PRESS)*.

Florida Sales Tax: For shipments to Florida destinations, please add 6% sales tax.

Shipping Charges: U.S. destinations add $2.50 for first book and $1.00 for each additional book. Foreign destinations add $6.00 for first book and $4.00 for each additional book. All U.S. orders shipped via Special Fourth Class Mail (Bookrate). All foreign orders shipped via surface carrier. Please allow three weeks for delivery.

Autographing Service: For autographed copies of books by Chevy Alden, please print the name of recipient(s) clearly and specify which books you want autographed.

Thanks for doing business with Tri-Pacer Press!